The Simple Beauty of the
UNEXPECTED

◉

MARCELO GLEISER

◉

# The Simple Beauty of the
# UNEXPECTED

A Natural Philosopher's
Quest for Trout and the Meaning
of Everything

ForeEdge

ForeEdge
An imprint of University Press of New England
www.upne.com
© 2016 Marcelo Gleiser
All rights reserved
Manufactured in the United States of America
Designed by Eric M. Brooks
Typeset in Arno Pro by Passumpsic Publishing

For permission to reproduce any of the material
in this book, contact Permissions, University Press of New
England, One Court Street, Suite 250, Lebanon NH 03766;
or visit www.upne.com

Library of Congress Cataloging-in-Publication Data
available upon request

5  4  3  2  1

To the trout I never caught,
and the equation I never solved

Yes, as everyone knows, meditation
and water are wedded for ever.

HERMAN MELVILLE, *MOBY-DICK*

◉

They do not step into the same rivers.
It is other and still other waters
that are flowing.

HERACLITUS, *FRAGMENTS*

◉

# CONTENTS

The Simple Beauty of the

UNEXPECTED

◉

Man is most nearly himself when he achieves
the seriousness of a child at play.

HERACLITUS

◉

# Prologue

## The Young Boy and the Sea

The boy inserted his fishing rod into a plastic pipe secured deep in the sand. The surf was low and the sun was already setting behind his back. Gone were the girls in scant bikinis and the muscular guys playing volleyball. Copacabana beach lay bare in front of him, a perfect, golden horseshoe. Here and there, older fishermen tried their luck along the beach, retired men in their sixties and seventies with little to do, their skin leathered from years under the tropical sun, beer bellies bursting out of their shorts. They all knew the persistent eleven-year-old who would come three or four times a week to the same spot with devout discipline. The routine was always the same: he'd string three hooks to the end of the line and carefully load each with a small piece of sardine. Then he would run to the water with the rod behind his back and cast the line as far as he could beyond the breaking surf. After placing the rod into the pipe, he'd sit down on the sand and wait. He paid little attention to the older men. Entranced, he shifted his gaze back and forth from the distant horizon to the tip of the rod. He didn't know then why he had to fish. But he knew he did. Alone.

Usually, he'd go home stinking of bait and empty-handed, or at best with a meager catch of a small catfish or a *cocoroca*, a bony relative of the sergeant fish common off the beaches of Rio. His older brothers would smirk, clamping their noses, amused at the boy's stubbornness. But not on that day. Two large silvery shadows darted fifty feet away, high on a wave. The boy retrieved his line quickly, hooked some fresh bait, and cast right behind where he had spotted the pair.

For ten minutes, nothing happened. Discouraged, the boy started to retrieve the line. Suddenly, he felt a strong tug. The bamboo rod bent in half with a fury he had never seen before. His arms turned rubbery. "It's a shark!" he yelled. "It's a shark!" Two older fishermen nearby dropped their rods and came closer. It had been years since someone had caught a shark there. The boy ran to the water's edge, holding on to the rod with all his might, trying to reel line in. But he could hardly turn the handle. "It's gonna snap! The line is gonna snap!" shouted one of the men. "Give up some line, boy! Let the fish run!" The boy, trembling head to toe, released the reel's lock. Line swished out as the fish tried to regain control of his destiny. The mighty predator had become prey to an even mightier predator, a stunned eleven-year-old boy. After some ten minutes of give and take, the boy finally reeled the fish ashore. It wasn't a shark. But it was big, bigger than anything he had ever caught or seen anyone catch at Copacabana beach. Silvery, flat on the sides with a large yellow tail fin; probably a young albacore, weighting about eight pounds, a beautiful creature to behold.

The older men circled the boy, amazed at the sight. Bursting with pride, the boy collected his equipment and tried to shove the fish headfirst into his bag. It wouldn't fit. The V-shaped tail stuck out as he walked the two blocks back home, pretending not to notice the looks of amazement from the passersby. He opened the door to his apartment and placed the fish on the kitchen counter. The cook, a large black woman in her fifties, came running in. "Lindaura, look what I caught for dinner tonight!" the boy said. "Grandpa is coming, right?" The cook eyed the fish with incredulity. "You caught this down by the beach?" The boy beamed. "I did. And no one helped me either. I wanna see who's gonna make fun of my fishing now."

It took me over thirty years to reconnect with that boy.

## You Give to the River and the River Gives Back to You

Life took me adrift and I forgot all about that young boy and his big fish. My head turned upward to the Universe, and I became a theoretical physicist, interested in questions that, not long ago, wouldn't have been considered scientific: How did the Universe come to be? What about all the matter that makes up stars, planets, and people? And life? Can we ever hope to understand how inanimate atoms first joined together to become living things, and then thinking brains? And if life took hold here, could it exist elsewhere? Could there be other thinking beings out there in the immensity of the cosmos? Even as a teenager, I marveled at the fact that such fundamental questions about existence could be answered in rational ways, without invoking supernatural agency. At the very least we could try to answer them, if not in their entirety, at least in part. The value, I realized, was in the trying, in being a participant in this continuous process of discovery we call science.

I now understand that those long afternoons of fishing and contemplation were a prelude to what was about to come. After all, fishing teaches us to be patient, tolerant, humble — key qualities needed in research. How often do fishermen go to the water with their rods, dreaming of the day's catch, only to come home empty-handed? Likewise, how often do scientists passionately explore an idea for days, weeks, months, years even, only to be forced to accept that it leads nowhere? Notwithstanding the frequent failures, and just as in fishing, they keep coming back, even if the odds for success are pretty low. The thrill is in beating the odds, occasionally landing a big fish or an idea that reveals something new about the world.

In fishing and in science we flirt with the elusive. We stare at the water, and sometimes we see a fish stir underneath the surface or even jump, betraying its presence. But the watery world is not our own, and we can only conjecture about what really goes on down there, polarized lenses and all. The line and the hook are our probes into this other realm, which we perceive only very imperfectly.

"Nature loves to hide," the Greek philosopher Heraclitus wrote some twenty-five centuries back.* We see very little of what really

---

*Or if this is not exactly what Heraclitus wrote — experts disagree, as there is no extant fragment with this quote — it is how his words have been distilled over the years.

goes on around us. Science is our probe into invisible realms, be it the world of the very small, of bacteria, of atoms, of elementary particles, or the world of the very large, of stars, galaxies, and even the Universe as a whole. We see these through our tools of exploration — our reality amplifiers — the telescopes, the microscopes, and the many other instruments of detection, the rod and line of the natural scientist. If we are persistent, once in a while we see Nature stir, even jump, revealing the simple beauty of the unexpected.

I left Brazil to pursue a PhD in England, moved to the United States as a postdoctoral fellow, got married, became a professor at Dartmouth, had three kids, got divorced, and lost twenty-four pounds in the process. Unless you fill up emotional loss with Twinkies and Big Macs, grief is a very efficient diet. A divorce is a small death, the death of a dream, of a relationship. You know you're inflicting pain upon the people you love the most, your children. You know you are violating their innocence with an irreversible sense of loss. It was the hardest decision of my life. But staying in the wrong relationship "for the sake of the children" would have been disastrous to everyone involved. You betray yourself, and by sacrificing who you are, you betray your children too. All you can do is be present, tell your story, and hope that when they grow up and have to deal with their own relationships, they will understand. And hopefully forgive.

I got lucky. I found a companion, married again, had two boys, and moved on. Although calling someone the love of your life may sound terribly naïve to the cynics out there, I can't imagine anything better than this in real life. And it's been nineteen years. Kari gave me many things, spiritual and material, that helped me turn my life around. Among them, and the one that matters the most for now, was a three-day fly-fishing course.

One sunny spring afternoon, we were walking along the Dartmouth Green in Hanover, New Hampshire, when I noticed a group of eight people quixotically brandishing long fishing rods in the air as if fighting invisible giants. By then, I had become a physics professor at Dartmouth College, holding the wonderfully titled Appleton Professorship of Natural Philosophy, a position that allowed me to teach hundreds of smart students and to wonder freely about the workings of Nature. Some call it work. I consider it a privilege.

I couldn't tear my eyes away from the swaying rods. Their dance made old neurons spark, stirring dormant memories back to life. A short man with a red baseball cap ran from one person to the other, yelling instructions, placing hands on the right spots, positioning bodies. Frustrated but always smiling, once in a while he would grab a rod and demonstrate how to cast. "Four beats, everyone: up goes the rod, back goes the line, forward goes the rod, forward goes the line! Get it? Lift it from one o'clock to eleven o'clock only! Strong wrists!" Rick Hammell is a true master of the art, although no one who knows him would call him a Zen kind of guy.

I stared longingly at the rod, at the fluorescent green fly line elegantly cutting through the air, darting some fifty yards forward. I saw a conductor with a very long baton standing by a mountain river, readying the elements for a performance. There was cause and effect, discipline mixed with excitement, man extending his reach to probe into a world beyond his own. A vision of the young boy, alone by the beach with his bamboo rod, flashed by. I was smitten.

"I know what I'm getting you for our anniversary," said my wife the soul reader at dinner, with a twinkle in her eye.

And so it was. I took Rick's fly-fishing course in the fall, bought the (expensive!) equipment, went out a couple of times, got frustrated with my ineptitude, swore I'd find time to practice my casting and, as so often happens in our lives, let it slide. I wasn't ready to give it what it takes. My work consumed me too much.

Learning how to fly-fish is no picnic. Handling the line, casting, reading the water, choosing what flies to use, making sure you don't slip on a rock and sink headfirst under the rushing current . . . As with all worthwhile things in life, you have to embrace it full-heartedly in order for it to work. The hands may move the rod, but it's the soul that moves the hands. Practice may give you technique; but gracefulness requires something else. Establishing this matter-spirit connection isn't easy. (If Rick heard me saying this he'd probably roll his eyes and hit me on the head with his rod. But heck, I'm a romantic.)

A couple of years passed without much fishing. Kari insisted.

"How come you're not going down to the river to fly-fish? C'mon, give it another try. Stop working so hard. Go have some fun!"

Kari was painfully right. Somehow, even after having taken the course, I'd let the long New England winters blur all memory of how much I wanted to reconnect with fishing. Spring would zip by amid mud and grayness, and then, by the time summer came, the usual family travels took over, friends and relatives came to visit, tantalizing work projects sprouted everywhere, there were books and essays to write, and before I knew it, summer gave way to fall and my dry flies stayed dry.

Well, not anymore! Something snapped inside me a few years back, when I decided to wake up at 5 o'clock one mid-August morning and walked down the path to what would become my fishing sanctuary on the Connecticut River. (Well, more a sanctuary than a fishing sanctuary, as fish are scarce there.) Why then

I'm not sure. Some emotional processes seem to have a life of their own, often grabbing you by surprise. I had just finished writing my second book in English, my research was going well, I wasn't teaching. I had room to let other emotional needs filter in. Besides, right outside our home the river beckoned . . . Within minutes I was knee-deep in water, waving my fly rod up and down, feeling like my imaginary conductor. Could the fish hear the music? I felt connected to an old and noble tradition going back centuries. And if we let go of the "fly" in fly-fishing — a more recent invention — we reach back to the dawn of humankind, as our hunter-gatherer ancestors devised means to catch food from the waters. Unfortunately, Rick's drilling kept stealing the moment: "Four beats, everyone: up goes the rod, back goes the line . . ."

I had to focus.

The night had been cool. Mist hovered above the water like a lover struggling to say goodbye. Dawn was a silvery blade slicing through the dark eastern sky.

I looked around. The clear water flowed past in a hurry. In the distance, the pink contours of Mount Ascutney emerged lazily from the darkness. I shuddered as my endorphin-bathed neurons flashed with primal contentment. Where had I been all these years? Sitting at the beach, the young boy looked at me and smiled. "It's time," he said. "Come."

Powers beyond my comprehension gathered around me. I was about to be baptized. The boy held my hand and led me a step further into the water. "Don't be afraid," he said. "I missed you, too." I smiled at my young self, reflected on the water. Time flows too fast, always faster.

Dunk, dunk, dunk. The boy baptized me and was gone. My life was about to change. I had entered the monastery.

The magic embraced me: the river, the fish, the quiet. Out

there, alone in the water, nothing mattered. All else dissipated like the morning mist, leaving a primal sense of contentment in its place. I could feel it inflating my chest like a party balloon.

I prepared my rod, a nine-foot, 6-weight, fitted with a green floating fly line and a 2x leader.* I like using bigger flies in the wide Connecticut, as fish, when they grace me with a visit, tend to be of considerable size. As the water was too warm for trout, I decided to go for bass — the smallmouth kind, common in these parts. Yellow poppers with red stripes seem to work well around here, although every fisherman who has tried to repeat his or her success knows that there are no real laws in fishing. Only suggestions. Like traffic lights after 10 PM in Rio.

I like to relate the surprising freedom we find in fishing to scientific research. There, too, are basic rules, such as the laws of Nature, which natural processes obey. But the thrill is to find freedom within these laws, or to go beyond them to discover the unexpected that lurks underneath the usual. What amazing designs can matter engender within the vastness of inner and outer space? What fabulous fish hide underneath that large boulder?

The Connecticut is quite dramatic around my sanctuary. The Wilder Dam, a few miles upstream, keeps it shallow, with the occasional deep pools here and there. Water flows fast in some

---

*The leader looks like a normal fishing line, light and transparent, to which we tie the fly. The difference is that it is tapered, becoming thinner toward its end. (This end part is often called the tippet.) The "2x" denotes the leader's relative thickness and thus strength. The 6-weight denotes the power of the rod, how thick it is. River fishing usually calls for rods from 1-weight to 6-weight, from small brooks to larger streams. Big ocean fish call for heavier-weight rods. The "floating fly line" is the thick colored line that propels the leader and the fly ahead. Different fishing conditions call for a floating, sinking, or partially floating fly line. The choice depends on the depth at which the fish are feeding and the current speed.

spots, and it can rise quite fast as well, when the floodgates are opened. I've had my share of semi-panicky late-summer afternoon escapes, threading knee-deep water while balancing on slippery rocks with one or two screaming kids in my arms. I learned very quickly to never underestimate a river.

I can only imagine what this magnificent waterway was like in the early 1800s, before dams and industrial mills. Salmon were so abundant that you could spear them from the shore. I once read they were in the hundreds of thousands then. Well, no salmon now . . . at least I haven't seen any. It would have to be a miraculous jumper to climb upstream. So, bass it is, always fun anyway.

A bald eagle flew overhead in perfect quiet, cutting through the air hardly flapping its wings. The clever thing knows all about Boyle's law of gases, the one relating pressure and volume. Knows it and uses it beautifully, gliding upward on an expanding hot-air bubble. Its kin have known this for much longer than we have. The eagle's mastery reminded me that there are many ways of knowing, and to remain humble. Knowledge about Nature need not be scientific or even human-based. We know so little. An arrogant scientist is like a peacock with lots of missing feathers and without a mirror to see himself. The same goes for every arrogant person, not just scientists. Arrogance wounds others and corrupts the self. As my grandfather used to say, "If you wear a hat bigger than your head, it covers your eyes."

Speaking of limitations, I hook a nice bass. Or rather, the bass hooks itself, leaping into the air to grab the popper before it hits the water, quite a spectacular sight. So spectacular, in fact, that I lose focus, let go of the line, give it too much slack, and the brave hunter manages to escape, twisting himself off the hook. Well done!

A few hours and many casts pass without any action. The sun

is up now, crowning a crystal-clear morning. What a piece of paradise this is. Why did it take me so long to come down here?

I know the answer well. Life carries me away from the river. I work too hard. I don't carve out enough time in my schedule to simply be. There are always research papers to write, meetings to go to, students to tend to, and, of course, my family. It shouldn't be a choice, but I make it so. It takes a while to learn how to chop time efficiently. It may not seem so at first, but the solution is quite simple: better time management and . . . a little less greed. How much do we need to achieve before we feel we've achieved enough? Looking back, I now realize that I wholly embraced fly-fishing only when I was ready to let go of some of this compulsion to do. At least, to do in the workplace. After all, to do is what we are here for. But we need stimulating diversity. To focus on a single activity is, to me at least, a bad (and boring) choice. I guess that in the dual world of foxes and hedgehogs, and with all due respect for the driven single-track hedgehogs, I choose to be a fox.*

Immersed in our busy lives, we find it hard to say, "I've done enough of this, now it's time to diversify, to take care of my inner self, to give a bit of my time back to myself." And as we know from all those New Year resolutions that remain unfulfilled, the hardest part is to act on them. But once I decided to act and give a bit of my time to the river, the river gave a bit of myself back to me. Yes, the river can do that; it can give yourself back to you, if you'll let it.

I hook another bass. Again I lose it. The reason, once more, was line slack. No surprise here . . . That's what happens when

---

*And speaking of foxes, I should mention that I don't "only" fly-fish. I'm also a devoted trail runner and obstacle racer, especially of Spartan races. The same way that financial managers are always telling me to diversify my investment portfolio, I'm a firm believer that in life we should diversify our doing portfolio.

you don't fish for almost a year. You get punished for your bad choices. Your casting suffers, your line management suffers. Fish get away.

No matter, the thrill was back. If fly-fishing were easy, it would have no mystique, nor would it serve as a metaphor for life.

I decided to get better. Like science, fishing needs mentorship. If you want to learn how to fly-fish, you need a mentor. If you want to become a research scientist, you need a PhD adviser. With very few exceptions, high-quality one-on-one instruction is the key to success in both. I had finished Rick's course and needed to go further. That first step into the monastery took me only as far as the vestibule. Unfortunately, no monks were waving at me, inviting me to take another step inward. I didn't really know where to go from there. But whenever my resolve started to waver, I could hear the boy, admonishing me: "Come on. Hurry up already! I've been waiting a long time for you to wake up."

If you enter the monastery in earnest, you don't walk out the same person.

As I was sitting at my desk, pondering how to find a mentor and what kind of fly-fishing experiences to pursue, I get an e-mail from Mark Hindmarsh, a colleague from Sussex University in England.

"Marcelo, we are organizing a workshop on classical field theory at Durham and would love for you to come. It's a small gathering, only UK physicists. You'll be our token American."

"Durham," I thought. "Right by wondrous Lake District. Lakes + rivers = trout . . ." I wrote back immediately. "I'm coming!"

And hence was born the idea of this book. Being a scientist, I have the privilege of traveling around the world to participate in all sorts of conferences, on topics ranging from the origin of the Universe to the origin of life to the meaning of the laws of Nature. What if, when going to conferences in different corners of the world, I also tried to fly-fish, and collected my experiences in a travel memoir? Sounded like a unique way of combining the two activities that give me a fuller sense of self. And writing about it would make it even more meaningful, especially if I could share this with others. What I didn't anticipate was the deep transformative power that my experiences would entail.

## A Note to the Reader

Please be aware that this is not the memoir of a master fly-fisherman, throwing out precious wisdom and victorious tales of amazing fishing trips, but that of an eager apprentice at the art of fly-fishing and the art of living.* If you know how to fly-fish, this book won't help improve your technique much. If you don't, perhaps you will learn something about this noble and humbling sport. Either way, my hope is that you will see the art of fly-fishing in ways you haven't before and, most importantly, learn a thing or two about the Universe and how science contributes to our search for meaning. Indeed, if you are one of those poor souls who couldn't care less about fishing, you may still carry on reading. In my life — and in this book — the fishing is mostly a conduit to the outer world of natural phenomena and the inner world of the self. While some engage in tea ceremonies or practice archery, and others bike across America or hike the Appalachian Trail, I fly-fish. The end goal is the same, to achieve some measure

---

*Even though I will use "fisherman" throughout this book, this is not at all to imply that this is a guy book or that fly-fishing is only for men. Quite the contrary, I've been put in my right place several times by master fly-fisherwomen. However, I'm sure the reader agrees that the gender-neutral "fisherperson" still sounds very unnatural.

of meaningfulness in your life. As I once told a journalist who had asked me — somewhat embarrassedly — what is the meaning of life, the meaning of life is to find meaning in life. This book is about my search for meaning.

# 1

# Cumbria, Lake District, UK

## On Unsolvable Mysteries

The journey had begun. I Googled "Fly-fishing Lake District," and bingo! Lots of places to go, beautiful, pristine rivers and lakes filled with trout and grayling, salmon too, depending when you go. I had hiked through parts of the Lake District with my girl-friend when I was a graduate student in England during the mid-eighties. My misty memories of the place were suffused with breathtakingly beautiful barren hills and warm flesh. Fishing couldn't have been further away from my mind then.

My next step was to choose a guide. Having but a few days to arrange everything, I e-mailed three names, all of guides based in the Lake District. The third on the list, Mr. Jeremy Lucas, was the one I decided to go with. Why? Well, Jeremy told me he would take me to the famous River Eden. I figured I couldn't pass up a river with such a name.

In the end, it was a very serendipitous choice. Jeremy turned out to be precisely the kind of guide-mentor I needed. He was then a member of the English fly-fishing team and had just gotten the silver medal at the most recent world championship. How about that? The clumsy apprentice would be learning from someone of such eminence! I was nervous, fearing I would greatly embarrass myself. I knew my casting left much room for improvement and that I had a lot to learn. I kept thinking of the famous words from the Buddha: "All beginnings are obscure." Fortunately, being a teacher helps you be a better student. I remembered that I too was a mentor and had students at all levels, from beginners to advanced. I was hoping that Jeremy would be as kind and patient with me as I try to be with beginners. You know that in order to learn you must *want* to learn. The world's greatest teacher couldn't teach someone who's not willing to be taught. And I wanted to learn all right. So I had that going for me.

⊙　　⊙　　⊙

Sometimes things do seem to happen for a reason. Some call these events happy coincidences, others call them the work of God, or of many gods, while others see them as manifestations of your karma. Being an agnostic (more on this later), I tend to like the happy-coincidence choice better, finding supernatural maneuvering a far-fetched hypothesis. In fact, the whole notion of a supernatural influence doesn't really make sense. After all, an "influence" denotes a physical occurrence or an event. And an occurrence is something that happens in the physical world through some kind of energy exchange. Any kind of energy exchange or force is very natural and requires a very natural cause. In other words, as soon as the supernatural becomes physical enough to be noticed or detected in some way, it can't remain supernatural anymore. A "supernatural influence" is an oxymoron. That said, there have been a few events in my life that defy logical explanation; at any rate, that defy any logical explanation that I could come up with. Although my position may sound somewhat shocking — especially coming from a scientist — to those who bet on our ability to explain everything, I'd argue that some things are unexplainable. In fact, I'll go further and argue that the unexplainable — to be distinguished from the not-yet-explained, which is the province of science — is unavoidable. And should be welcomed.

We are surrounded by mystery, by what we don't know and, more dramatically, by what we can't know. Hence my metaphor of the Island of Knowledge, which I elaborated in a recent book and briefly review here:* if our accumulated knowledge of the world makes up an island, the island grows as we learn more. (It may also occasionally shrink, as we discard an erroneous theory or explanation.) As with every island, this one is also surrounded by an ocean, in this case the ocean of the unknown. However — and here is the twist — as the island grows, so do the shores of our ignorance, the boundary between the known and the unknown. In other words, new knowledge generates new unknowns. Unless

---

*The Island of Knowledge: The Limits of Science and the Search for Meaning (Basic Books, 2014).

we stop asking questions about Nature, there is no possible end to our search. Furthermore, scattered along the ocean of the unknown are regions of unknowables, questions beyond the reach of scientific inquiry. We will have more to say about those later on.

Powerful as they are, our brains are limited, as are the tools of scientific inquiry, the machines we use to collect data about the world. Every measuring device has a range and a set precision. Telescopes can "see" only so far — that is, they can collect light from sources up to a certain distance. Whatever lies beyond their reach can't be seen, even if it's as real as what is seen. The same applies to microscopes, of course. Tiny things may escape detection, even if they are there, as real as the things we can see with the naked eye. If we continue into the world of subatomic particles, the smallest entities that exist, how far we can probe into the heart of matter depends on the machines we can build. Particle accelerators, such as the Large Hadron Collider at CERN, the European Organization for Nuclear Research, can probe matter only up to a certain limit. Whatever exists below that limit goes undetected. We may increase the accuracy of our machines and thus probe smaller distances, but we can't do this indefinitely, up to "zero" distance. There is no perfectly accurate, all-seeing measurement. We are permanently myopic to some fraction of what exists.

Therefore, we must conclude that this ever-growing body of knowledge called science cannot explain all there is for the simple reason that we won't ever know all there is to explain. How could we possibly know all the questions to ask? To presume that we can know all there is to know only shows how supremely arrogant some people can be. It also flies against all that we have learned about how science generates knowledge.

Some may consider my task of exposing the limits of science to be dangerously defeatist, as if I were "feeding the enemy." (I've been accused of that.) But that is surely not the case. To understand the limitations of science is not the same as labeling it as weak or exposing it to the criticism of antiscience groups, such as Bible literalists. It is, in fact, liberating to those who consider it, as it frees science from the burden of being godlike, all-knowing and all-powerful. It protects its integrity in a time when so many claims from scientists get inflated beyond their validity, either by those making them (they should know better) or by the media. Cases in point: statements that we understand the physical mechanisms behind the Big Bang, which we do not, or that life is ubiquitous in the Universe, which we know not. Scientists should be quite careful about what they say and how they say it, as their pronouncements carry weight in the social sphere. Furthermore, and this is a key point for us, why should we *want* to know everything? Imagine how sad it would be if, one day, we arrived at the end of knowledge. With no more questions to ask, our creativity would be stifled, our fire within extinguished. That, to me, would be incomparably worse than embracing doubt as the unavoidable partner of a curious mind. Science remains our most effective tool to explore the world in its myriad manifestations. However, we shouldn't lose sight of the fact that it is a human invention and that, as such, it does have limitations. Every system of knowledge is fallible. It needs to be in order to evolve. Failure compels change. Besides, we don't want reason to invade every corner of our existence. Some mysteries can be solved by reason, and others just can't.

## Immortally Beautiful Open Spaces

It was all arranged. Jeremy would pick me up on Saturday at 9:30 AM at Collingwood College, where the conference participants were housed. I could hardly contain my excitement. But first, of course, there was the conference, the official reason I was in Durham. Classical field theory . . . three words that everyone knows separately but that sound very bizarre and mysterious when put together. "Classical" relates to works — in classical music, or the classics of world literature, or classical Greece and Rome — that earned distinction; more than that, immortality. A work is a classic when, in spite of its old age, it's relevant today and will still be when we are gone: classic is the opposite of ephemeral. People will be listening to Beethoven, Mahler, and the Beatles for a very long time. What books being written today will be read (or the futuristic equivalent of reading) two hundred years from now? Although I have my list, I doubt it will overlap much with those of others. It takes time for a work to become a classic.

"Field," on the other hand, usually means open spaces where things grow, or where ball games are played. In Portuguese, the language I grew up speaking, *campo* — the word for field — can also mean countryside.

So, is a "classical field theorist" someone who comes up with theories of immortally beautiful open spaces? Sounds like poetry. In reality, though, classical field theorists, even if some may be poets or see Nature as a poem written without words, seek to express themselves in other ways. In classical physics — to distinguish it from quantum physics, which studies the world of atoms and subatomic particles — a "field" is a sort of extension in space of a source of some kind. For example, a body creates a temperature field around it that falls quickly with distance: moving away from the body, different points in space have different temperatures. The collection of these measurements, in principle including every point in space, is the temperature field around the body.

If the body moves about, so does its temperature field, creating a "time-varying" field.

Another familiar field is the one any concentration of mass creates, the gravitational field. Every massive body attracts every other massive body. We all know that a rock suspended in air falls down if we let go. In reality, rock and Earth attract one another with equal (and opposite) strength, but the much more massive Earth "wins," and it's the rock that moves. (A body's mass is a measure of its *inertia*, the resistance it offers to a change in its state of motion.) We can visualize a field surrounding both rock and Earth, extending their gravitational pull out into empty space, like someone wearing strong perfume. The concept of field alleviates the mystery of the Newtonian "action at a distance," the fact that objects may attract one another (or repel one another) even without touching. We know that if we want to make a ball move we need to kick it or throw it. But the Sun makes the Earth move about it without any touching. Why is that? With the concept of the field, we can picture a ghostly presence in space interacting with the Earth and causing its motion.

Fields are usually strongest near the source, weakening with distance. A field that actually grew in intensity away from its source would probably violate a few natural laws.* As Isaac Newton showed in 1686, the gravitational attraction of any body with

---

*There are, of course, exceptions, most notably in the world of particle physics. For example, we now picture protons and neutrons as being composed of three smaller particles called quarks. Quarks exert a force on each other mediated by another class of particles called gluons. The gluons have an associated field, the gluonic field. Trying to separate quarks from each other puts a stress on the gluonic field, somewhat like stretching a rubber band. This generates an elastic force that grows in intensity with the distance between the quarks, equivalent to a field that grows with distance.

mass, from a rock to a person to the Sun, weakens with the square of the distance from it, a very precise statement that caused quite a sensation in Europe at the time. Newton's new physics changed the way people saw the world around them. Likewise, the concept of field, a later, nineteenth-century invention due mostly to Michael Faraday and James Clerk Maxwell, changed the way we picture the nature of physical reality, what philosophers call ontology. Instead of the Newtonian particles moving about because of the action of forces, after Faraday and Maxwell we would picture particles moving under the influence of fields. Instead of an empty arena where things happened, an inert background, space became an entity filled with stuff, a concrete player in the nature of physical reality. No longer atoms moving in the Void, as the pre-Socratic philosophers Leucippus and Democritus had conjectured around 400 BCE but fields filling up the void and making things move about.

We are literally surrounded by fields: the gravitational field created by the Earth and by everything else that has mass; the electromagnetic fields from countless kinds of electromagnetic radiation, from the visible light from the Sun and lightbulbs to the invisible radio waves beamed from hundreds of FM and AM antennas, the ultra high-frequency radio waves from cell phones, the infrared heat from our bodies and every warm object . . . the list is very long.

In fact, for physicists like me who deal with the inner structure of matter and the cosmos, *everything* is a field of some sort, coming in two types: matter fields and force fields. Matter fields comprise essentially all the matter that makes up our bodies and the stuff around us: rocks, air, water, refrigerators, iPods, stars. Think of atoms and their constituent electrons, protons, and neutrons. Each one of these matter particles has an associated field that defines its identity: the electron has its field, and so does the proton. You can picture each particle as a chunky excitation of a field, a kind of energy knot that moves about in space and interacts with other energy knots. A suggestive (and only that) image is that of small waves moving atop the surface of a swimming pool, crashing against one another. The obvious difference is that water waves tend to disperse after colliding with other waves or obstacles, while particles tend to keep their shape.*

The other types of fields are force fields, which describe how matter particles interact with one another. Think of two people talking to each other. The people are the matter fields: they interact via words, their "communication field." This field establishes a verbal link between them. According to modern physics, force fields are the communication fields of matter fields. And they come in several kinds (four so far). Pushing the analogy a bit further: just as people in different places speak different languages, different particles feel different forces. An electron, for example, has a tiny mass and a negative electric charge. So it attracts other electrons gravitationally (very weakly, owing to its tiny mass) and repels them electrically (much more strongly). These two force fields are needed to describe interactions between electrons. Back to our analogy, words are not the only way people interact. There are other "communication fields" that are used, such as looks and body movement. To fully describe how people interact, you need to include all possible "communication fields."

A chunk of matter made of (electrically neutral) atoms, and

---

*As we will soon see, there are exceptions to both cases: there are water waves that don't disperse after crashing with other waves, and particles that do break into smaller bits after colliding with other particles. Some particles may also spontaneously decay (disintegrate) into a bunch of smaller particles, like a bomb into bits of shrapnel. Nature is much more interesting than our strict classifications.

hence with a mass and no net electric charge, gravitationally attracts other chunks of matter made of (electrically neutral) atoms. Consider, as an illustration, you and Earth, two sizable chunks of matter with vastly different masses. Each has an associated gravitational field. If you are far away from Earth — say, on Mars — the gravitational pull Earth exerts over you (and you over Earth) is so small as to be negligible. But as you come closer, the two fields overlap more strongly, and you feel a harder pull toward Earth. The gravitational field is the messenger, telling how the two chunks of matter should attract each other. More generally, force fields such as gravity or electromagnetism are the links, the bonds between matter fields. And they add up. Right now not just Earth but also the Sun, the Andromeda galaxy, the rings of Saturn, all the objects around you, your children away in school or your parents at work, your worst enemy, your secret love, the world's most horrendous criminal as well as its most enlightened being, are pulling at you gravitationally. And you are pulling back at them. Fortunately, the pull weakens with (the square of the) distance and is sensitive to the amount of mass. So the attractions from faraway things and light objects don't exert much of a pull on you. Otherwise, moving about would be a very complicated process, as massive objects would tend to blob together into a tangled mess.

Let's pause for a second to contemplate how fluid this portrayal of the inner workings of Nature is, connecting everything and everyone. As the nineteenth-century naturalist John Muir wrote in *My First Summer in the Sierra*, "When we try to pick out anything by itself, we find it hitched to everything else in the universe." Physics tells us how this universal hitching works.

Field theorists describe the world combining mathematics with the results of experiments that measure how particles of matter — themselves excitations of fields like ripples on the surface of a lake — interact with each other via force fields. To a greater or lesser extent, everything influences everything else. Isolation is an abstraction, at best a useful approximation.

Reality is a weblike structure of mutually interdependent influences, of which we perceive very little.

We may now move on from fields to "classical." Again, the meaning is a bit unusual. The distinction here is with "quantum," the physics that describes the world of the very small, of atoms and particles. Since quantum theory is a more recent invention — dating from the early twentieth century, as opposed to the seventeenth — the older pre-quantum description became the "classical" theory. Fair enough, as it all started with Galileo and Newton, whose works certainly deserve the "classical" accolade.

Given that science is a description of the natural world at all scales, it's best to consider classical and quantum theories not as radically different (even though they are in their formulation) but as complementary. Quanta are the smallest units of matter and energy, just as the cent is the smallest unit of currency. Every financial transaction, no matter how big or small, is performed in multiples of a cent. The electron is a quantum of the electron field; the photon is a quantum of the electromagnetic field, of which visible light is a familiar manifestation. When physicists refer to "elementary particles of matter" they mean the smallest possible bits of stuff, those that can't be broken into even smaller ones.*

---

*However, we must be careful about statements like this, given that what was unbreakable yesterday may be breakable tomorrow with more powerful tools. In the language of particle physics, particles that look elementary today may actually be a composite — that is, made of even smaller particles. The recently discovered Higgs boson, for example, is a good candidate for being a composite. In other words,

The essential point is that classical fields emerge from quantum fields when there are huge numbers of quanta present. Imagine a sandy beach from afar. It appears nice and smooth — classical — even though it's made of many "quanta," the grains of sand, each a discrete unit.* This analogy also tells us that the quantum description is mostly appropriate at short distances, when the graininess of matter can be perceived and is relevant. Needless to say, we humans have a very classical perspective of reality: the quantum world emerges only when probed by highly sensitive experiments designed to "see" the inner graininess of matter. Bringing it home, classical field theory describes the behavior of fields composed of many quanta: back to the beach analogy, it focuses on the motions of the sand dunes and not on that of the individual grains of sand.

## Solitude and Solitons

The meeting at Durham was mostly about lumps of matter (and thus of energy, as the two are related by the famous $E = mc^2$ relation) made up of so many quanta that the equations of classical field theory are the ones that best describe their properties. These lumps have many names — even just "lumps," which to me sounds a bit too cancerous. Thankfully, the most common is "soliton." I love this word, soliton. It's used to represent a remarkable and somewhat freakish phenomenon, a gathering of matter that

---

a hundred years from now the list of elementary particles of matter almost certainly will be different from today's.

*The quote marks are a reminder that grains of sand are not, properly speaking, quanta. It's amazing to think that each of them contains over a trillion billion silicon and oxygen atoms. But the image contrasting smooth and grainy is certainly suggestive of the distinction between continuous and quantum-like.

doesn't dissipate, retaining its shape as it moves about in space with apparently nothing to hold it together. The word evokes solitude, or solitary, originating from the "solitary wave" that the young Scottish engineer John Scott Russell first witnessed in 1834. On a sunny August day, Russell was comparing the relative efficiency of horses versus steam for drawing boats along the Union Canal, near Edinburgh, when the rope connected to the boat snapped and "the boat suddenly stopped — not so the mass of water in the channel which it had put in motion; it accumulated round the prow of the vessel in a state of violent agitation, then suddenly leaving it behind, rolled forward with great velocity, assuming the form of a large solitary elevation, a rounded, smooth and well defined heap of water, which continued its course along the channel without change of form or diminution of speed" (Russell, 1844). Amazed, Russell rushed to pursue the watery apparition. He "followed it on horseback, and overtook it still rolling on at a rate of some eight or nine miles an hour, preserving its original figure some thirty feet long and a foot to a foot and a half in height," until the wave got lost along further windings of the channel. The serendipitous experience was life-transforming for Russell. From then on, he spent much of his time researching the properties of solitary waves in water tanks. Among his many findings, the strangest is the fact that solitary waves cross each other without "change of any kind," that is, without losing their shapes, as if they were ghosts. This is very counterintuitive. A wave usually moves for a while and then spreads out and dissipates. If it crashes with another wave, or with a big rock, it loses its shape. Not so with solitons. They stay on and on, defying common sense. What holds them together? In the case of solitary waves in water, a balance between dispersive and gathering tendencies conspires to keep the wave's shape as it moves along. Even though details vary, most solitons result from a similar balancing act. A Taoist would say that the soliton illustrates the Tao, the wholeness achieved as the yin of dispersion is balanced dynamically by the yang of gathering tendencies — the attractive interactions among its constituents. To a physicist, a soliton is a bundle of particles interacting with each other so as to behave as a single, non-changing entity: a macro-particle, so to speak, the many behaving as one. And to a Taoist physicist, a soliton is the embodiment of permanence from change, of being from becoming.

The word "soliton" — the *on* stresses that the object is like a particle, like the *on* in electron or proton — comes from "solitary"; solitary leads to solitude; and solitude leads to . . . fishing. To me, fishing is best done alone. A lone ritual. It's about giving yourself time with yourself, making an offering to the river goddess out there, the one awaiting by the water while we're indoors, wondering why we don't get out more.

Fishing is about you, the water, and the fish. It's about surrendering to the moment, disconnecting from all else. It's about losing yourself in a timeless realm through action and intense focus. It's about achieving being from becoming. One could almost call this *Zen and the Art of Fly-Fishing*. But let's not.

All the things I love to do — physics, fly-fishing, guitar playing, writing, trail running — are solitary activities. No, I'm not the stereotypical schizoid, asocial, nerdy scientist type. At least I don't think so. I love people, and, like most, I love to be loved. This search for loneliness is different. It's not, actually, a search for loneliness. It's not about hiding away. It's a search for Nature, for the companionship it provides. Through physics I try to decipher its secrets, through fly-fishing I engage viscerally with it, through music I attempt to re-create its harmonies, and through writing I attempt to engage with my experiences and register them in a more permanent way. This manifold devotion, this search for different ways to connect with something bigger than I am, can only be called love. Einstein called it the experience of the mysterious — "the cosmic religious feeling" — to him the most significant we could have, the awe we feel as we contemplate Creation. (By "Creation" with a capital C I mean the totality of Nature.) In my view, it is the purest form of spirituality, the manifold experience of our profound connection with the cosmos. From Nature we came, in Nature we are, to Nature we go. Maybe this could be my epitaph.

## Pattern Seeking

The talk went really well. My British hosts were nice enough to let me speak for an hour, even though everyone else had a half-hour slot. They must have felt they owed me that, since I was the only one coming from afar. Little did they know how much I wanted to be there.

My subject was "oscillons," yet another kind of *ons* (and a name proudly invented by me, along with other colleagues I wasn't aware of). These, though, were not found running along river canals. They were found in my computer, as solutions to the equations describing how certain fields evolve in space and time.* Oscillons are like solitons in that they too are localized lumps of

---

*Two Russian physicists, I. L. Bogolubsky and V. L. Makhankov, found oscillons two decades before me, calling the objects pulsons in 1974. I was unaware of their work (as were, apparently, most high-energy physicists) until I went through the necessary bibliography check before submitting my paper to the editors of the *Physical Review*. I was somewhat disappointed at not being the first, but was also happy that my research on oscillons was validated. I spent the following years exploring the properties of oscillons well beyond these initial results, in particular showing how oscillon-like structures appear in a wide variety of physical theories.

energy. But they have an extra feature that, at least to my mind, makes them truly remarkable, even more than the watery apparition that Russell followed on horseback. Unlike solitons, which always keep their shape as they move about, oscillons, as their name suggests, oscillate. They are structures that live for a very long time despite all odds and expectations to the contrary. In Nature, things that oscillate usually do so only for a while. Either they stop because of friction, like a swing or a leaf in the wind, or they just spread out, like a wave that is unable to retain its shape as it advances in space. Not so with oscillons: they just keep on oscillating for a long time, much longer that what seems reasonable.

Here's a good way to picture an oscillon: imagine you throw a stone in a pond. You see vibrations, concentric waves propagating outward from the point of contact with the water until they eventually die away in the distance or hit a boundary. If the water had the right properties to sustain an oscillon, you would see the central oscillation continue on and on, without concentric waves moving outward. It's almost spooky! In Nature, when a disturbance is created, its energy usually dissipates in some way. That's what concentric waves are doing after the stone hits the water — they carry the energy away from the point of impact. Likewise, a pendulum clock without a mechanism would not swing for very long. Can you picture the scene? You throw a rock on a pond, and against all common sense, the region where the rock hits the water keeps bobbing up and down without shedding its energy in concentric waves. Anyone's immediate reaction would be, "Okay, what's the trick?"

Unfortunately, oscillons don't form in water, at least not yet. They need special kinds of fields to manifest themselves. But they exist out there, no question about it. In 1995, a year after I published my first paper on oscillons, a group from the University of Texas in Austin also published a paper on "oscillons" — coincidentally, they chose the same name, even though they were examining a very different kind of phenomenon! The Texan oscillons were formed from a bunch of little glass beads inside a cylinder that oscillated up and down with a tunable frequency. The authors noticed that persistent oscillating patterns formed by the glass beads emerged for certain frequencies of the cylinder: oscillons in vibrating grains . . . It seems to me that their oscillons, and the many types that have been found since 1994 in different physical systems both real and hypothetical, are manifestations of the same universal phenomenon, a dynamical balance between the natural dispersive tendencies of wavelike phenomena and an attractive force that wants to bundle matter together, as gravity does. The trick about the first simple oscillons I found (subsequently, my research group and colleagues elsewhere found more complicated kinds) is that this attractive tendency is efficient only when oscillations are large and thus hard to excite naturally. (As in a swing, large oscillations require large pushes.) They need a lot of energy to show up, together with a property called "nonlinearity": a small initial disturbance may cause an unexpectedly large response. Like someone in a bad mood who explodes out of all proportion.

Nature often repeats patterns at different scales. Think of the spirals of hurricanes, of stirred cream in coffee, of spiral galaxies, and of snail shells. Or think how trees, rivers, and our own arteries and veins (and nerves, too, for that matter) branch out, as do our bodies from torso to arms and legs and then to fingers and toes. Spiraling and branching are two very common patterns, seen on many different living and nonliving things. They result from the joint action of the two key principles of natural design: energy efficiency and optimization. Every natural phenomenon follows the path of least energy: a rock falls in a straight line because it is the shortest path between two points and hence the most economical. Soap bubbles and balloons attain a round shape so as to minimize surface tension. Snowflakes acquire a six-pointed shape owing to the geometrical structure of water molecules. But each flake is the unique result of an optimization process sensitive to the weather conditions in its immediate surroundings — moisture, temperature, atmospheric pressure. As water turns to ice, it sheds its extra energy in the most efficient way possible, resulting in snowflakes that are never exactly alike, each encoding the story of its birth.

## Leaders, Followers, and Outsiders

But my talk was not about spirals and snowflakes. I reviewed many results on oscillons I had obtained alone and with my collaborators over the years, emphasizing recent ones. The morning session over, the conference concluded with a lunch at Collingwood College, where we lodged. Ah, the ever-ghastly British college food . . . I guess certain traditions never die. It reminded me of my PhD days at King's College and the cafeteria special, a pasty-gray meat stew with sulfurous boiled Brussels sprouts. Even after more than two decades I can still smell it. As a personal vendetta, I now cook a wicked good dish of stir-fried Brussels sprouts with grilled tofu and mango curry, and have avoided eating meat for years, pasty gray or otherwise.

Food woes apart, it was a wonderful time spent with colleagues. Where else but in science conferences do you find a group of people openly sharing their knowledge, not caring much about the way they look, and with almost total disregard for authority? Graduate students, postdoctoral fellows, professors, famous professors, we all sit together to talk about our models and assumptions, eagerly trying to learn and criticize each other's works. Outside the conference halls the subject usually drifts toward the funding situation, or who is working where, and to meaty stuff on so-and-so's divorce or tenure case. Deep down, we are just a bunch of gossipers with a weakness for math and a fondness for Nature.

The picture above, of course, is a bit too rosy. It's not quite true that there is widespread equality in science. Sure, that's what's in the charts: being right is what counts, not authority; winning a Nobel Prize doesn't turn you automatically into a prophet of divine truth. However, as with any human endeavor, in science too there are leaders, followers, and outsiders. The leaders are the trendsetters, the ones who seem to have all the good ideas — or at least the most attractive ones — and, almost as important, know how to broadcast them. Often, again as in any human endeavor, the broadcasting is more attractive than the idea itself. The followers add refinements here and there, enriching the research and sustaining the leaders without questioning their wisdom. The outsiders are the ones who refuse to jump onto the bandwagon. They search for their own problems to work on, even if this usually complicates their professional lives: when you go your own way it's harder to get a job, to get papers published, to get your work referenced by others, and to get your research funded.

When you think you have a nice original idea, it's hard to see everyone else working on something completely different and — to your mind — useless or just plain boring. You feel in your gut that you are onto something, that it must be pursued at all costs, even if your colleagues or adviser think you're wasting your time. And if things don't always quite work out the way you had hoped, if your wonderful idea doesn't pan out — and believe me, that happens more often than not — that's okay. At least you know you had the intellectual integrity and courage to pursue your own intuition. A cynic might say that you're just being foolish, trying to follow the archetype of the lone genius whose visionary work is vindicated in the end. That should never be the reason. An outsider must follow her intuition in humbleness, not knowing where it will take her. A fisherman never knows if he is going to catch anything. Still, you revisit your idea and go back to the river again and again, hope renewed each time. You do it because you wouldn't be yourself otherwise. You do it because you don't want to look back at your life and regret the choices you made when it's too late to do something about it. There's no fishing in heaven. And if there were, it would be quite boring, as you'd endlessly catch all the fish you want. What's the fun in that?

## The Simple Beauty of the Unexpected

With the conference over, I decided to take a walk around Durham. With its magnificent castle and well-preserved eleventh-century Gothic cathedral, the river Weir snaking around its contours and ancient stone bridges crossing it here and there, Durham ranks as a true medieval jewel. Miraculously, the heavy morning rain had stopped, and strong winds had started to clear the sky.

A public footpath meanders along the river. I approached it through a narrow alleyway just beneath the castle. A huge sycamore bowed ceremoniously over the dark green water. I paused to appreciate the view, infused with a deep sense of peace. A cloud of mayflies wobbled just above the current, joyfully celebrating their twenty-four-hour existence.* Suddenly, out of the depths,

---

*The scientific name for mayflies says it all: they belong to the order of Ephemeroptera, from the Greek *ephemeros* = "short lived" and *pteron* = "wing."

a salmon leaped some three feet into the air, swallowed one of them, and dived back with a noisy splash. The fish must have been at least six pounds, maybe more. I just stood there, motionless, mouth agape.

If there are such things as signs, this was one. Nature had just sent me a message; at least that's how I saw it, which is what matters. Few moments in my life had been more meaningful. A cozy warmth spread across my chest, as I experienced a kind of revelatory awakening. I had just witnessed the simple beauty of the unexpected. "You need to get out into the wilderness more often. You're missing the magic," said a voice in my head. This time, I was listening.

The sheer power of the fish was staggering. What was a salmon doing so close to town anyway? I found out later that this is the season of peak salmon activity in Northern England, when such visions are not uncommon. What a gift!

Was it an omen? Of course it was! Only a fool, blind, sad rationalist would wave away something like this, dismissing it as a mere coincidence. When an event is meaningful it becomes more than a mere coincidence. I'm not saying that some supreme supernatural power or some river spirit planted the message just for me. That would be nonsensical and hopelessly self-centered. The salmon jumped, and I happened to be right there to see it. Why take away from the simple beauty of what had just happened, attributing it to an invisible and elusive conductor? What should be worshipped here is not some invisible, unknowable magic hand but the serendipity of the event, the emotional impact it had on me. The salmon's timeline and my own overlapped for a few brief seconds of pure and absolute bliss. There is no need to bring anyone or anything else into the picture.

The incident left me exhilarated. Next morning, I was to take off to the Cumbrian Mountains with Jeremy. But there were still a few hours, which I could use to sort e-mail accumulated over a couple of days, call my family, and have dinner. Although Collingwood is a good twenty-minute walk from the center of Durham, I hurried out again to get dinner in town. Anything to avoid the college fare. The night was surprisingly clear — was this really Northern England? The moon, three-quarters full, floated above the cathedral like a broken silver shield. I wondered if the salmon had any idea of the moon, if it had ever leapt out of the water to try and swallow it whole, thinking it a giant bright moth. No, it probably knew exactly what it was: not a satellite world bound to Earth by gravity — that's our scientific version of the moon — but a night giver of light, a clock of sorts guiding migratory excursions up and down the river. That salmon will stay with me forever.


## Belief

Saturday had finally arrived. The first thing I did when I woke up was to open the curtains. It was sunny! Having only one day to fish, I really didn't want to spend it under heavy rain. I would, mind you. But sunny was definitely better.

The second thing I did was to eat a huge English breakfast, ingesting enough fat and carbohydrates to keep me going for a few days. It actually tasted great, even as it insulted my granola-tamed insides.

Jeremy arrived at 9:30 sharp, naturally. A gentle man with kind eyes, he had the demeanor of someone who had seen pain and found peace. That was promising. Most of us have see pain, but few find peace.

"We have about an hour until we reach our destination," he said. "I wanted to take you to the wonderful river Eden, but with the heavy rains the waters are dark and cloudy, and the fishing won't be as good."

I was expelled from Paradise before even getting there. In truth, though, I knew it didn't matter where I'd end up going. I didn't really care about catching fish. All I wanted was the experience of wading upstream with an expert guide, someone who would help me take another step inside the monastery.

"Well, it's okay," I said. "What I want is the instruction, you know, casting, line handling and retrieving . . ."

"Oh, don't you worry about that, Marcelo, you'll have plenty of it."

"Where are we going?"

"I'm taking you to one of the wildest spots in England, a lake atop some hills in the North Pennines, at about two thousand feet. I find it very beautiful, and it's teeming with wild brown trout."

Remote lake? Teeming with wild trout? Sounded wonderful.

We started to climb through Northumberland, passing countless fields — of the pasture kind — postcard-perfect English rolling hills crisscrossed by ancient stone walls. Here and there, puffy black and white sheep polka-dotted the vast expanse of land.

We crossed the inspiring medieval village of Middleton-in-Teesdale, right by the very fishable river Tee. I looked longingly at the fast-moving waters. Jeremy read my thoughts.

"Yes, I'm glad I'm taking you up to the lake. The water does look clouded up," he said.

I agreed absentmindedly, not bothered at all by the cloudiness of the water.

"By the way, Marcelo, what was your conference on?"

"Oh, I'm a theoretical physicist."

"How amazing! I'm a theoretical chemist."

"Really?"

"Yes, got my PhD in 1982, working on orbital theory in London."

"In London? Me too! I got my PhD from King's College."

"How remarkable is that? So did I!"

So, my fly-fishing mentor was a PhD scientist. And with a degree from my alma mater . . . I couldn't have made this up.

There are many fly-fishing guides in Northern England, as I realized when I used the Web to find one. But here I was, riding along with a fellow Nature lover who, like me, studied it scientifically and enjoyed it through his fishing. In his case, though, science had stayed in the past, at least as a focus of active research.

"I have fly-fished for forty years, and there is nothing I'd rather do. Except, of course, my writing."

"Wait. You also write?"

"Oh, yes, novels and nonfiction. Why? Do you?"

"I do indeed!"

"I guess we really are kindred spirits."

"No question about it," I enthusiastically agreed. Jeremy and I have remained in touch ever since, even if sporadically. He recently sent me a copy of his masterly *Tactical Fly Fishing*, a guide for advanced and competitive fly-fishermen, a kind of graduate text for fly-fishing.

Coincidences like these, that quickly establish a meaningful connection with someone who, moments earlier, wasn't even part of your life, inspire many to believe in some sort of supernatural explanation for things. We feel special when life offers these sudden opportunities for spiritual growth. There is gratitude and even a measure of primal fear, an in-your-face confirmation of our lack of control over so much of what happens to us, sometimes good, sometimes bad. In our powerless condition, we feel a need to find something to grab on to, to find some kind of order behind the random succession of events that dictate so much of our lives. How else can we keep our sanity? To many, there is only one way: to believe in a controlling power, a supernatural over- seer concerned with each individual's well-being. This power can be (a very busy) God, but it can also be astrology or some other cosmic source of influence based on some vague pseudoscientific principle. And so people become believers, searching for invisible connections between events in their lives, often missing the sim- ple beauty of the unexpected.

Belief is not the exclusive province of the religious. Secular sci- entists also believe, albeit in a different way. For an illustration, we go back to the dramatic early days of quantum mechanics, when many scientists desperately tried to find order in the midst of en- suing chaos. In the 1920s, Einstein, Planck, Schrödinger, and oth- ers struggled to find an underlying explanation for the apparent randomness of the atomic world that dozens of laboratory ex- periments were exposing in astonishing succession. These men believed in the classical worldview established since the days of Galileo and Newton, a view of Nature grounded on determinism, on processes taking place continuously, responding to simple cause-and-effect relations. There was an underlying logic, a clock- work precision that allowed for a sense of control in figuring out how Nature worked. The world of atoms, it was becoming ever more apparent, did not seem to conform to such rules of good behavior. And yet, in trying to make sense of experimental data, these men ended up becoming the pioneers of the new quantum worldview, putting forward revolutionary ideas that led to the way we look at physical reality today. Their angst, their reluctance to accept a worldview that was so contrary to their beliefs, gives new meaning to the expression "reality bites."

The quantum realm they helped unveil is strangely alien to

ours: electrons jump discontinuously from atomic orbit to atomic orbit, more like kids going up and down stairs than billiard balls rolling up and down a slide. According to classical physics, the atom itself should have been unstable, as electrons and protons strongly attract each other with nothing to hold them apart. This being the case, why didn't electrons fall over the protons? What kept them zooming around in their orbits? Furthermore, calculations in quantum physics can predict only the *probability* that an event will occur — for example, the probability that an electron will be found at this or that location at a certain moment. Gone was the triumphant deterministic power of classical physics, where planetary orbits can be computed with such extreme precision that we can predict exactly the date and time of a solar eclipse thousands of years into the future. Quantum reality was vastly different from the world of steam engines, cannons, mills, cars, and hydroelectric power stations, machines that obey mechanistic laws and can be constructed and trusted to follow them. In the world of the very small, reality is defiant, mysterious, unrelated to what we are used to. Schrödinger, when confronted with the fact that his own theory created a worldview he refused to accept, experienced a sort of Dr. Frankenstein nightmare and regretted having had anything to do with quantum mechanics. Einstein openly challenged his colleagues on the completeness of quantum physics.

"This is only half the story," he would claim. "Quantum mechanics is an incomplete theory, built out of necessity to explain unusual experimental results. It simply cannot be the final word on how atoms operate. There has to be an underlying order behind all these probabilities and mysterious behavior."

"And why, Professor Einstein, do you think this way? What makes you so sure?"

"Because the world cannot be so disordered. The Old One doesn't play dice."*

This is belief, science-style. It's not about faith in God, it's about faith in a worldview, in the way the world ought to be. Einstein's famous dictum expresses his belief that Nature must function in a deterministic way, that it simply couldn't be so disordered at its essence. His "Old One" is his metaphorical "God," Nature following deterministic rules, a Nature whose mysteries are accessible — at least partially — to human reason.

During the early twentieth century, the new quantum worldview was essentially forced down the throat of a large number of scientists, contradicting their long-held philosophical value system. What can one do when confronted with the new and unexpected? There are two options: embrace the new and accept that the old views are in need of revision; or stick with the old view and *believe* that there is a deeper way to explain the new based on the old. This was Einstein's choice. (And Schrödinger's, Planck's, de Broglie's . . .) The key point is to recognize that in the absence of conclusive data, either option entails belief. Even so, we must be mindful of the fundamental difference between a scientific and a traditional religious belief: dogma. In science, no belief can be sustained in the face of mounting opposing evidence. There is no arguing against well-established fact. There may be resistance to change, but eventually wrong ideas are abandoned. (It may take the old guard to die first, as Planck once wittily remarked.) In religion, since direct evidence is elusive or irrelevant, belief is *always* a viable choice.

---

*This imaginary dialogue was paraphrased, of course. But it does reflect Einstein's feelings toward quantum mechanics. Niels Bohr, another giant of twentieth-century physics, famously yelled back at Einstein: "Einstein, stop telling God what to do!"

There is, however, an interesting complication in science, in particular in modern high-energy physics and cosmology. Some theories may not be testable or capable of being proved wrong. Like a zombie that never dies, it is possible to have a theory of physical processes that can always be redefined to escape laboratory or observational tests. Every time evidence points against that theory, its parameters are adjusted so that the newly revised theory makes predictions beyond what can be tested. It's like a stairway to heaven, with steps being added as we climb up.

Let's go on a short digression to see how this plays out for real in theoretical physics. The reader who isn't interested can jump to the next section break. I hope you will read on, though, if only to explore how belief plays a role in the very practice of science.

⊚   ⊚   ⊚

Currently, there are two major ideas in fundamental theoretical physics that fit this situation: supersymmetry and the multiverse. Starting with the less bizarre: supersymmetry is a hypothetical symmetry of Nature whereby every ordinary particle of matter has a supersymmetric companion, somewhat like a mirror image, but with a tricky mirror that changes a few things. The theory would help resolve a series of conundrums we currently face in high-energy physics, at the cost of doubling up the number of "elementary" particles. The key prediction from supersymmetric theories is that at least the lightest of its particles should be stable: that is, the particle doesn't decay into another lighter particle or particles.* In other words, according to supersymmetry, there

---

*We mentioned that unstable particles might spontaneously decay into lighter ones. This is true for most particles, in fact. An isolated neutron, for example, decays into a proton, an electron, and the more exotic "electron anti-neutrino" in about fifteen minutes.

should exist a new particle out there as stable as an electron. And if there is a new particle, we should be able to detect it, in accelerators like the Large Hadron Collider (LHC), for example, or in detectors that study particles that rain down from the skies known as cosmic rays. But despite very aggressive searches during the past decades, this hypothetical particle hasn't shown up. It also hasn't shown up on the first run from the LHC, where the Higgs boson was found, the particle responsible for giving mass to all other particles of matter.

In April 2015, scientists restarted the LHC with twice the energy of its previous runs: the higher the collision energy, the more massive the particles that are produced. Most defenders of supersymmetry (or SUSY) believe that this will be it, that when the new data are properly analyzed (around the time this book is released in 2016, or shortly thereafter), the new particle will finally be detected. If it is detected, excellent, we will enter a new era of very exciting high-energy physics. But what if it isn't? Will supersymmetry be abandoned as a possible theory of Nature?

This is where things get interesting. I predict a split in the high-energy physics community. The strains are already showing. Some will throw in the towel and concede that SUSY is not there. Others won't, arguing that the absence of evidence is not evidence of absence, that SUSY could be realized at energies much higher than what our current or future machines can probe. For these physicists, SUSY as an explanation for physical reality will become an article of faith, an untestable hypothesis — in short, an intangible belief. And an untestable hypothesis, even if temporarily useful in aiding extrapolation and conjecture, has little permanent scientific value if the predictions it inspires can't ever be verified. It might be a convenient or even beautiful idea; but how can we know if it has anything to do with Nature? I can't wait for the LHC data to be released so that we get to the bottom of this.*

The multiverse is an extension of current theories of cosmology, the part of physics that studies the Universe as a whole. If you thought the Universe was big, think again! The essential idea is that our Universe is not all there is, but part of a much larger entity called the multiverse. Different theories call for different types of multiverse. In general, the multiverse is a collection of a huge number of universes, each with its own laws of Nature. Ours happens to be the one where physical laws allow for stars to grow old so that at least one planet may support living matter. What's in the other universes — that is, the kind of physical laws that define them — depends on the theory. If you think that this whole story is somewhat vague, it's because it is — starting with the notion of a "physical law." If you check the literature, even just Wikipedia, you will quickly see that there is no one list of physical laws that scientists agree upon. There are some obvious ones, like the law of conservation of energy and electric charge; but things go haywire rather quickly. So, when physicists talk about universes with different physical laws, especially in the context of the multiverse, we must be careful to distill what they mean. The answer will depend on the type of theory that generates the multiverse.

A popular example comes from the so-called "string theory landscape," an expression that needs a lot of unpacking. "String theory" refers to a set of theories that make a radical change in the way we picture the world (philosophers like to call this an ontological shift): instead of having elementary particles as the basic

---

*Not only because I am very interested in knowing if SUSY exists or not, but also because I have a bet with Gordon Kane, a colleague from the University of Michigan and one of the world experts in supersymmetry. At stake is a fifteen-year-old bottle of Macallan.

blocks of matter, the little Lego bricks of all there is, the fundamental entities are now tiny vibrating tubes of pretty much pure energy. Just like when you make different sounds in string instruments by changing the length of the vibrating string (usually by pressing down on it with a finger) so that the shorter the string the higher the pitch (that is, the frequency) of the sound, fundamental strings can vibrate with different frequencies, each corresponding to a different elementary particle. With strings you get this all-in-one picture where a single fundamental entity can reproduce the whole family of elementary particles, a very economical description.

Unfortunately, life is not so simple. For string theories to make physical sense — that is, for them to give rise to reasonable predictions about the real world — they have to have two essential properties: first, they need to be supersymmetric; second, they need to be formulated in nine spatial dimensions, six more than what we see around us. We discussed the issues with supersymmetry before. If SUSY is not a property of Nature, string theory — superstring theory, actually — is pretty much dead. And if we do live in a space with nine dimensions, we need to understand what happened to the extra six. As a quick reminder, you can picture a "dimension" as a way to represent a direction of space where movement is possible. So, a straight line is a one-dimensional space, since an object can move only up and down the line. To know someone's position along the line or along a circle all you need is a single number (the distance from a reference point, or the angle from a reference angle). A plane surface (like a tabletop) is a two-dimensional space, since an object can move along two directions. The space we exist in is three-dimensional: we can move along the floor and also jump up and down. If there are extra dimensions, we don't see them. Neither do our microscopes or our particle accelerators, like the LHC. You may be wondering, why talk about microscopes and accelerators when discussing dimensions? The reason is that these extra dimensions can be very tiny; for example, they could be wrapped around themselves in a kind of six-dimensional ball of minuscule radius. To picture this, imagine a simpler situation, a stretched-out rope. From very far away, a rope looks like a line, and hence like a one-dimensional space. But as you move closer to it you see that it has an "extra" dimension, its thickness, which we can represent as a tiny circle attached to each point along the line. (The stretched rope is, at least approximately, a very long cylinder.) The extra dimension, being so tiny, can be seen only under close scrutiny, or amplification.

Microscopes and, more dramatically, particle accelerators are reality amplifiers that allow us to see small or invisible things, including small extra dimensions.* So far, we haven't seen any indication of extra dimensions. However, this is not conclusive, given that they can be much smaller than what we can probe; so small, indeed, that we could never hope to probe them. We've seen this before. If we can't ever probe extra dimensions directly, can we be certain that they exist? "Certain" is not usually the right word to use in science. If observations confirm some of the predictions inspired by extra dimensions, what we could say is that we'd have indirect evidence that they exist. For example, we should expect to see some particles with properties that relate to the size of the extra-dimensional space. But this and other predictions are secondary. The really amazing prediction from superstring theory is that our whole Universe is a consequence of these extra dimen-

---

*More precisely, you don't "see" an extra dimension. A particle that goes that way seems to disappear into thin air and is accounted for as "missing energy" in the detector. (No, ghosts are not objects going into extra dimensions of space!)

sions: the properties we observe here, the kinds of particles, their masses and electric charges, etc. — everything that exists — is a consequence of the underlying string theory!

No wonder I fell in love with strings in my graduate school years, and wrote a PhD thesis about the cosmology of universes with more than three spatial dimensions. How beautiful and elegant it would be if our Universe sprouted from a multidimensional space, all its properties determined by the geometry of the extra dimensions. Isn't this the most grandiose goal of science, to predict *everything*?* Science as oracle, no secrets left out.

We can trace this kind of expectation to Einstein and, a long time before him, to the philosopher Plato in ancient Greece.

The complication is that since a space with six dimensions can be twisted, folded, and cut in all sort of ways (imagine doing it to a ball made of Play-Doh), each geometric possibility (or configuration) corresponds to a different three-dimensional universe: the geometric rules of the extra dimensions translate into the physical rules of the three-dimensional universe. Counting all the particular kinds of six-dimensional spaces, we arrive at a ridiculously huge number of possible universes, something like $10^{500}$. This is the somewhat sad state of string theory today: its original appeal as *the* theory that would predict our Universe from a unique geometric arrangement of the extra six dimensions degenerated into a plethora of possible universes, each with a different set of physical laws. How are we to pick the "right" one — that is, the one that represents our Universe — out of this crazy mess? What are the selection criteria we should be using?

---

*I'm not talking about why you play or don't play guitar, or what you will do tomorrow. "Everything" here refers to the properties of the physical Universe, in particular those of the elementary particles of matter and their interactions.

That's the angst of superstring theory. From a theory that would have our Universe as the *only* possible solution, it became the theory where all kinds of universes are possible. But physicists are a resourceful bunch, and soon a tentative resolution was proposed: instead of stressing over finding *the* solution that would be our Universe, turn the whole thing around and breathe with relief that ours is a possibility among many others. The values of the physical parameters in our Universe (masses and charges of particles, how fast it expands, etc.) happen by chance to be the ones that allow for us (that is, living creatures) to be here. Knowing the properties of our Universe, we can narrow down the values of the physical parameters that make universes like (or very similar to) ours possible. But we can't predict these values, which was the noble goal of string theory earlier on.

Even if we go along with this humbler approach to string theory and consider the possibility of varying physical parameters across the multiverse landscape, each set of values corresponding to a different universe, we still can't ever probe other universes directly. *The multiverse is not a directly verifiable physical entity.* Its existence is an unknowable. Remember that what we can measure of the universe is confined inside a bubble of information: like a fish in a bowl, we can't know what's outside it. Even if the cosmic information bubble is truly huge, with a radius of some forty-six billion light-years, it's still finite. Other universes are by necessity outside it.

A useful analogy is the horizon we see at the beach, where sky and ocean seem to touch. We know that there is more ocean beyond the horizon (unless, like the ancients, you believe that the horizon marks the end of the world), but we can't see it. I often tried, as a young boy, sitting patiently by my rod waiting for the fish to bite. I'd spy large ships approaching from the distance, knowing that since I saw their masts before their hulls, they had to be coming from beyond the horizon. Likewise with the Universe: all information that we gather about the cosmos comes from different kinds of light or, more precisely, of electromagnetic radiation. That's what telescopes are — buckets that collect different kinds of radiation. We may image objects directly (visible light), or we may detect their infrared or ultraviolet or X-ray or radio emissions. The cosmic horizon comes from the fact that, like you, the Universe has a birth date. Although details of what happened back then remain fuzzy (more on this later), we do know that time as we know it started ticking some 13.8 billion years ago. This means that light, traveling at the ridiculous speed of 186,000 miles per second, could cover only a finite distance in this finite amount of time. This distance determines the cosmic horizon, the radius of our information bubble. Anything beyond the bubble remains necessarily out of reach. Just as there is more ocean beyond the horizon at the beach, there probably is more Universe beyond our cosmic horizon. But strictly speaking, we can't be sure of it (as we can be of the ocean beyond the horizon), even if it is a very reasonable extrapolation.

The best that we can do is to hope for indirect evidence of the multiverse. A far-fetched idea is that if other universes exist "out there," they may have collided with ours in the past. Colliding universes can be a very dramatic event; obviously, if any such collision happened and we survived to tell the tale, it would have been more like a cosmic brushing-past than a head-on hit. In any case, such encounters could leave a signal within our cosmic horizon, tiny waves in our information bubble. These waves would have very peculiar shapes that would be imprinted, like patterns on a fabric, on the radiation spread across space known as the cosmic microwave background. Although searches so far have not produced anything very promising, it's interesting to think that the stuff in our Universe may carry scars of old cosmic battles.

⊚   ⊚   ⊚

The hard question then is what to do with such acts of faith from scientists. When data — the lifeblood of science — are absent for too long, or indefinitely, things get muddled. Being the creation of humans, of people with dreams and aspirations, science isn't perfect. Its practice does reflect its practitioners, especially as we catch it in the making, where the boundaries of knowledge are fuzzy. My worry is not that people may get so carried away by an idea, bewitched by its appeal to such an extent that their ability to judge is compromised. Although passion and critical thinking are not always compatible, scientific knowledge is constructed in such a way as to be immune, at least in the long run, to such passions. If we can get data, things eventually settle. The problem starts when we can't get data to disprove a long-standing hypothesis. The danger then is to let passion blind us to such an extent that we are not able to let go, even when everything tells us to. We become slaves to faith.

I understand that hope is a key engine behind our creativity, and that it's not always clear when to let go of an idea. (Or of a person, for that matter, often with very painful consequences: "She will love me one day, I just have to keep showing how much I love her . . .")

There is, however, a point when persistence becomes a tragic fantasy. The physicist Albert Michelson, the first American to win the Nobel prize, died believing in the existence of the luminiferous aether — the medium that supposedly supported the propagation of light waves — even if his own experiment from decades earlier was the key evidence against it.* There are many examples like his. And how devastating is it to see the one you love — and that you hope will love you back — in the arms of another? Change the "one you love" in the previous sentence to "your theory of how Nature works" and you get the picture. It hurts to be wrong; it hurts much more when what is wrong is not just an idea, but a whole worldview. As every fisherman knows, if the fish don't bite after hours and hours of trying, it's best to pack up and go home. At least with love and fishing we can try again.

## The Witch of Copacabana

Who hasn't experienced strange things in life, events that defy explanation? Instead of digressing into a long argument as to why such strange events are just coincidences, and that we are the ones who attach meaning to them because we are meaning-seeking creatures, I will tell you a bizarre story, a hair-raising event that remains — at least to me — mystifyingly unexplainable.

When I was seventeen, I spent long hours studying for the grueling university entrance exams in Rio, a terrifying rite of passage for most Brazilian teenagers fortunate enough to go to high school. It is no exaggeration to say that these exams make SATs look like fifth-grade quizzes. When I took them, you had to answer questions on topics ranging from algebra to world history to genetics, competing with tens of thousands of smart kids for a few hundred coveted spots at the best universities. Contrary to what happens in the United States, where teenagers tend to see college

---

*Unlike water or sound waves, which need a material medium to propagate (water for water waves; air for sound waves), light (electromagnetic radiation) travels in the void or vacuum.

as a means to move away from parental control (while many parents see it as a much-needed break from their kids' moody and rebellious affect), in Brazil college students mostly remain in the city where they grew up and in their parents' home. Apart from key differences in family dynamics, this also means that Brazilian students have severely limited choices of good schools to aim for. In my case, there were only two or three choices. To make things worse, my two older brothers were ridiculously successful students, both getting into the school they wanted to, at the top of the rankings. The pressure, real and self-inflicted, was brutal.

My parents loved hosting dinner parties; we used to have at least one a week. This was 1976, a time when many members of the Portuguese upper class had fled to Brazil, refugees from the Socialist government that took power after the "Revolução dos Cravos" of April 25, 1974—the Carnation Revolution. The bright red flower became the symbol of the revolt after a soldier put one in the barrel of his rifle as the victorious troops paraded the streets of Lisbon to the enthusiastic acclaim of cheering crowds.

My father was a dentist by profession, although his true talents lay elsewhere. Everyone but himself loved him, a charming, well-read man with a wonderful musical ear and a passion for gardening and antiques. Thanks to an extensive network of mutual friends, the newly arrived Portuguese flocked in droves to his dental practice, happy to find a European soul in tropical Rio. Politically, he leaned to the right, which made the refugees even happier.

One day at lunch, my father announced that we were soon to have a very important guest for dinner, Senhor João Rosas, Portugal's former minister of justice (somewhat like the US attorney general), together with some friends from Lisbon. My stepmother Léa, who enjoyed parties even more than my father, smacked her lips and started to think of an appropriate menu. "The pièce de résistance should be *bacalhau espiritual*," she said, already tasting it. A delicious soufflé-like dish made with dozens of eggs, "spiritual cod" is as light as a spirit would be if one could be trapped in food. "Let's show these Portuguese that we can cook their own food better than they can." The revenge of the colonized! (Comparing the British and the Portuguese culinary traditions, I'd guess that this revenge is much easier to achieve in the United States.)

After much commotion and preparation, the day of the dinner party finally arrived. By 8 PM all the guests were chatting away in the living room. Ever the gracious host, my father approached Senhor Rosas, a diminutive man with a noble curving nose and a floral silk handkerchief tucked into his navy blazer's left pocket.

"João, can I offer you a drink?"

"Whisky is fine, Izaac."

In seconds, my father was back with the whisky. A meticulous man, he kept all alcoholic beverages in an old shower cabinet refurbished as a liquor closet. The minister took a sip and paused, his eyes opening wide.

"Ó, Izaac, I'm very sorry, but this is not whisky."

"What? What do you mean?"

"It's tea. Yes, certainly tea. Nice tea, mind you, but just tea."

"Impossible! Let me try." My father grabbed the crystal tumbler from Senhor Rosas's hand. "My God! It *is* tea! I am terribly sorry! How could this . . . I'll be right back, João, please forgive me a moment."

My father rushed to the beverage closet and tried the whisky from the bottle he had served Senhor João. Tea. He eyed the other three or four opened whisky bottles on the shelf. Tea in all of them. Horrified, he moved to the cognac bottles. Also tea. Every single bottle of amber-colored spirit was filled with tea. The good dentist of Copacabana almost spontaneously combusted.

He rushed to the kitchen, where Maria, our cook, was muttering something over the cod soufflé, no doubt a prayer to conjure Yemanjá or some other spirit to the room. She was a small black lady in her late fifties, with pitch-dark beady eyes. A white turban perpetually covered her head, no matter what outfit she wore. We all knew what that meant: Maria was a high priestess of the Macumba, a syncretic religious practice widespread in Brazil, mixing African black magic and fetishism with elements of Catholicism. On Mondays, the day of the souls, countless candles illuminate crossroads around the country, many with offerings of dead black chicken, cheap cigars, and half-empty bottles of *cachaça*, together with pictures of loved or hated ones. Macumba rituals involve a lot of drinking and chanting. The ritual induces the "channelers" into a trancelike state so they can "receive" the spirits of the dead. Once they are possessed, their back arches, their eyes roll, and their motions become jerky as they give advice in matters of the heart in otherworldly guttural voices that sound like Yoda with a hangover. Highly regarded in the Macumba tradition, channelers are those rare souls with an open door to the beyond, the conduits to the world of the dead. Maria was one of them.

"Maria!" my father yelled, startling the lady from her drunken stupor. "Did you drink everything in the closet?"

"Almost everything, yes, sir," answered Maria, unabashed, barely lifting her eyes from the food plate. My father was livid.

"Tomorrow morning I want you to pack your things and get out!"

Maria turned toward my father. We were used to her glassy eyes, and to how she would hover above the huge pot of black beans for hours, oscillating slowly from side to side in a trancelike state. This time, however, her eyes looked different. They weren't just glassy. They were shooting angry sparks of light. My father took a step back. I was standing right behind him and saw his hand go slowly into his front left pocket, where he always kept a head of garlic. In my father's world, evil could strike at any time. It was a perpetual war.

"I will go, Doctor, but something will happen to this house. Just you wait!"

A curse! Maria, the Macumba high priestess, had put a *curse* on our home. This was bad. Remembering his VIP guest, my father retreated to the party, with a sealed bottle of Chivas Regal in his hand.

Next morning, Maria called me into the kitchen. She had packed her bags and needed help getting them downstairs. I tried hard to avoid her gaze but couldn't. Her eyes were still sparking. She grabbed me by the shoulders and stared right into my soul.

"You, boy, you have *corpo fechado*. Nothing will do you harm."

Petrified, I managed to thank her awkwardly while twisting myself from her grip. *Corpo fechado*, literally "closed body," meant a kind of spiritual shield that protected a person from evil.

For the next few days, my father carefully nursed the rue shrub he kept with other plants on the veranda. In Brazil, many believe that the rue plant is a sort of chlorophyllous evil barometer that shrivels away when a person of evil intent enters a house or gives someone in the family the evil eye. Fortunately, the shrub looked healthy. My parents hired a new cook, after making sure she didn't own a white turban, and the family moved on to other things.

About a month after the whisky incident, I was studying in my room when I felt a cold chill run down my spine. This was odd, because it was November in Rio, and the temperature was around ninety degrees Fahrenheit. I tried concentrating on my math problems but couldn't. I felt an uncontrollable urge to walk down the corridor, toward the dining room. Our rococo-style dining table was flanked on both ends by furniture containing fine crystal. Behind my father's seat at the head of the table was a closet with glass doors and three glass shelves, where my parents stored the "too good to use" wineglasses made of Bohemian crystal, some with golden rims, others beautifully etched with floral patterns. At the opposite end of the table was a brass beverage trolley, with a top glass shelf covered with crystal bottles filled with port, sherry, and liqueurs of all colors, each labeled with a small silver necklace. It was all for show, as no one really drank in my family.

I was standing by the dining table in a strange sort of daze when something, maybe a subtle noise, made me turn toward the closet. At that very moment, the top shelf broke in half, and all the heavy glasses came crashing down onto the second shelf, which in turn collapsed onto the first shelf in a horrifying waterfall of shattering crystal. Dozens of priceless antique glasses were instantly destroyed. I hardly had time to blink, when another cracking noise made me turn toward the trolley at the other end of the table. In a flash, the top shelf collapsed, taking all the crystal bottles to the floor with it. The noise was deafening. Shards of glass flew everywhere. I was paralyzed. The new cook came running from the kitchen and crossed herself. She packed her things and vanished that same night, never to be seen again.

Shaking uncontrollably, I phoned my father at his office. "It's the curse, dad. She did it! Everything crashed, right in front of me. The closet and the trolley, practically at the same time!"

"Don't touch anything! I'm coming home!"

Poor father. The event tore open the thin veil that, in his mind, separated the world of the living from the world of the dead. "Yo no creo en las brujas pero que las hay, las hay," he used to say: "I don't believe in witches, but they surely exist." After this day, how could he not believe?*

Apparently, we had hired the queen witch of Macumba herself. I was stunned. How could something like this happen? Coincidence? Sure, if it had been only the closet *or* the trolley. There was tension on the shelves, they were overloaded, years of exposure to tropical humidity had rotted the wood pins supporting them . . . But *both* closet and trolley practically at the same time? And in a region of the world where there are no earthquakes or even slight tremors? Could it possibly have been some sort of resonance effect, the specific sound frequencies of the breaking crystal inducing more crystal breaking? Highly unlikely. A supersonic blast wave from a jet flying nearby? Nah. Let's accept what happened for what it was: a very bizarre occurrence of spooky synchronicity. Any rational explanation simply doesn't add up.

The disaster couldn't have been more attuned to the curse, as it happened to wineglasses and liquor bottles. And what about the shivers I felt, the urge to walk to the dining table? There's no question that I witnessed a very unnatural occurrence. Had Maria used me as her conduit? Was that grabbing of my arms, that fiery staring into my eyes some kind of hypnotic technique? Did I break everything under some sort of sleepwalking trance and couldn't remember a thing? Not very probable. I'm not prone to hypnosis or to sleepwalking. And as far as I know, I don't suffer from multiple personality disorder and forget what my other self had been up to. I actually loved, even if from a distance, the crystal glasses from Bohemia. The fact is that both the closet and the trolley collapsed in near-perfect synchrony. The curse had been

---

*This saying, even in Brazil, is always quoted in Spanish. My father claimed it came from the poet Federico García Lorca, although I was unable to verify this.

fulfilled. The witch of Copacabana had the final laugh. *Yo no creo en las brujas pero que las hay, las hay* . . . This is one mystery I'll never figure out. Perhaps it's better this way. Not everything must be explained, not every question must have an answer. Life would be quite boring otherwise. A bit of the unexplained is good, keeping us a little unsettled.

## Reason, Faith, and the Incompleteness of Knowledge

I try to picture myself as a seventeen-year-old, going through this experience. I was terrified. In fact, I still am, at least a bit. If not terrified, at least mystified by the whole thing. Any explanation that I can come up with challenges what I consider to be "normal." If I broke everything and can't remember it because I was in some kind of hypnotic trance, then it means that there are dimensions to my existence beyond my control. That's pretty scary. If I didn't do it and it happened through some supernatural magic, then my whole worldview is in bad need of revision. If it happened through some perfectly natural cause, and I—or anyone else I told this story to—can't figure out what it was, it means that there is much more to reality than we know. This last option is by far the best. It sustains my hope in our ability to comprehend the world at least in part, even when faced with what is apparently incomprehensible. After all, isn't modern science a tool to comprehend what is so different, so distant from our immediate reality? Doesn't science probe into the mysterious, inching its way into understanding, explaining the unknown with the knowable?*

---

*In contrast, traditional religions attempt to explain the unknown with the unknowable, simultaneously feeding on the mystery and feeding it back.

That is science's main goal, to embrace the unknown, trying to make sense of it. However, the acquisition of knowledge, by its very nature, is an endless pursuit. As our Island of Knowledge grows, so do the shores of our ignorance, the boundaries between the known and the unknown. The more we know, the more we discover we don't know. As the history of science teaches us over and over, new discoveries, new tools, have the power to change entire worldviews. Think of how we pictured the sky before and after the telescope; or how we understood life before and after the microscope. These instruments changed the way we see Nature and our place in it: if before we saw humans as God's creation living in the static center of the cosmos, we now see ourselves as evolved primates living on a small blue planet among trillions of other worlds in our galaxy.

A key ingredient of the island metaphor is not only that are we surrounded by unknowns, but that some of these unknowns are unknowable: there are well-posed questions that science cannot address. There are natural phenomena that we can't explain, possibly ever. This declaration may fly in the face of those who believe in a sort of scientific triumphalism, that science can conquer it all. Well, it can't. We scientists must have the integrity and clarity of vision to know where to draw the line, acknowledging what we can't handle, at least through the scientific method as we now understand it. Here are a few current examples of unknowables, from the cosmic to the cognitive:

- We can't know what's beyond our cosmic horizon, the bubble of information defined by the distance light has traveled since the Big Bang, around forty-six billion light-years.
- We can't explain in deterministic fashion the essential randomness that takes over at the quantum level, having to

accept that the possible outcomes of a measure are probabilistic.

⊕ We can't construct a self-referential logical system that is closed, in the sense that every statement within that system can be proved, as the Austrian mathematician Kurt Gödel showed with his incompleteness theorems.

⊕ A computer cannot include itself in a simulation; it is thus impossible in principle to simulate the Universe as a whole, since the simulation would by necessity include itself.

⊕ Humans may be cognitively impaired in understanding their own consciousness, a problem known in the cognitive neurosciences and related philosophical discussions as the "hard problem of consciousness."

There is clearly a lot that I'm leaving out of each of the aforementioned unknowables, as I covered them in my previous book. Of course, the notion of an unknowable must be considered with care, given that what may seem unknowable today may become knowable tomorrow. However, for the examples above, this unknowable-to-knowable transition would require very fundamental revolutions in our understanding of physical reality, such as the following:

⊕ Faster-than-light travel needs to be possible; or inter-universe-connecting traversable wormholes need to exist and be stable.*

⊕ A whole new way of thinking about quantum physics needs

*Wormholes are somewhat like tunnels in space-time that are mathematically possible in Einstein's theory of relativity. The reader may think of them as an imaginary network of subway tunnels across space, an image Arthur C. Clarke explored beautifully in *2001: A Space Odyssey*. Unfortunately, current models of wormholes call for very exotic physics, bordering on the implausible.

to be developed, one that explains the apparent randomness of results from measurements. Previous efforts based on local "hidden variable" theories don't work.

- A new, self-contained logical structure for mathematics would have to be created.
- New concepts in computing, where a machine can simulate itself, would need to emerge, an apparent impossibility.
- The human subjective sense of self that baffles the cognitive neurosciences would have to be explained as an emergent property of complex neuronal activity.

In one way or another, all these unknowables share one common characteristic, in that they require achieving some sort of total knowledge, an all-encompassing explanation of a given physical and/or biological system from within:

- To know the Universe as a whole, without being able to step outside it.
- To explain all possible results in a quantum system, including the relative probabilistic weight for each of them.
- To have a complete mathematics.
- To have a complete simulation, one that includes itself.
- To have the brain explain itself in its entirety.

Such barriers to an all-encompassing knowledge remind me of my intellectual heroes, Jorge Luis Borges and Isaiah Berlin. Both argued against the possibility of achieving such a goal, calling on us to be humble about what we can and cannot do. Both explained how futile and misguided such efforts are, as they lead to a false sense of conviction that we have abilities beyond our means. They also carry a powerful destructive force, when implemented socially: oppressive ideologies stem from the misguided belief that their foundational value system is superior to others.

In "The Library of Babel," Borges tells the story of the keepers of a fictional library that contains all books ever written and that will ever be written. The main challenge for these librarians was to find the catalog of catalogs, the one that would include all the information found in the infinite library. Surely, any such catalog would need another one that would include it in its listings. The structure is like that of an infinite onion, with more and more outer layers, ad infinitum. There can't be a complete catalog of all knowledge, since a catalog cannot include itself; another one would be needed to include it, and so on. Borges is playing with the idea of the infinite, making it clear that it should remain an idea; any attempt to convert it into reality is bound to fail. We humans are these amazing creatures who, although finite in physical extension and reasoning power, are capable of conceiving the infinite. We can imagine it, we can calculate with it, and we can even represent it graphically, as Italian Renaissance genius Brunelleschi showed with his science of perspective, using projective geometry to bring the infinite to the two-dimensional plane of a canvas. Amazing as we are, let us not be fooled into claiming the *reality* of the infinite, lest we choose to live a lie. The infinite may exist — for example, the Universe may be infinite in spatial extent — but its reality is unknowable: we can't ever be sure whether space is infinite or not. No measurement can prove it beyond a doubt. We may believe in the infinite, but not that we can grasp it in any sort of concrete way.

Berlin has repeatedly insisted on the impossibility of any kind of absolute or total knowledge, calling it the "Ionian Fallacy":*

---

*"Ionian" here refers to the group of early Greek philosophers from the region known as Ionia, on the western coast of Turkey. As far back as 650 BCE, the Ionians looked for a unifying explanation for the material world, believing that all that exists is made of a single kind of matter. Thales, the founding father of the group, believed all was water.

"A sentence of the form 'Everything consists of . . .' or 'Everything is . . .' or 'Nothing is . . .' unless it is empirical . . . states nothing, since a proposition which cannot be significantly denied or doubted can offer us no information." In one of his last addresses, to the University of Toronto on the occasion of receiving an honorary doctorate, Berlin expanded his views to the sociopolitical sphere:

> The idea that to all genuine questions there can be only one true answer is a very old philosophical notion. The great Athenian philosophers, Jews and Christians, the thinkers of the Renaissance and the Paris of Louis XIV, the French radical reformers of the eighteenth century, the revolutionaries of the nineteenth—however much they differed about what the answer was or how to discover it (and bloody wars were fought over this) — were all convinced that they knew the answer, and that only human vice and stupidity could obstruct its realization.
>
> This is the idea of which I spoke, and what I wish to tell you is that it is false. Not only because the solutions given by different schools of social thought differ, and none can be demonstrated by rational methods—but for an even deeper reason. The central values by which most men have lived, in a great many lands at a great many times — these values, almost if not entirely universal, are not always harmonious with each other.*

As soon as you believe that your values are superior to those of others, you create a differential that can easily be turned into an oppressive ideology. This has happened over and over in history

---

*"A Message to the 21st Century," *New York Review of Books*, October 23, 2014.

with tragic consequences, as Berlin reminds us. And it has happened in the sciences as well, albeit in more subtle ways.

There is no question that in science we must strive for simplicity and for theories that have as wide a range of applicability as possible. In the eyes of a physicist, a "beautiful" theory is one that uses a minimum amount of conceptual input for a maximal amount of explanatory power. By extension, the most beautiful of theories would be the one that explains it all, where by "all" we mean the behavior of matter in its most diverse manifestations. (As we've seen before, unified theories in physics are not designed to deal with complex human choices or with predicting the weather two weeks in advance or whether you will win the lottery.) Under this prism, the idea of unification is the "true answer" to the question of Nature: at its very essence, Nature must be simple; all forces are but different manifestations of a single one.*

It is understandable why such a theory is seductive. I spent a good fraction of my career pursuing it with many colleagues; but in the light of our previous arguments concerning the incompleteness of knowledge, we are forced to abandon such enterprise as being more of a belief system than a plausible scientific goal: monotheism turned scientific. Given our limited grasp of reality, we should learn to accept that we will never be certain whether we have achieved a complete knowledge of all material interactions. For example, a new force of Nature might be discovered

---

*In his recent book *A Beautiful Question: Finding Nature's Deep Design*, the Nobel laureate physicist Frank Wilczek offers a moving and inspiring "meditation" on how beauty acts as a guiding principle toward unveiling Nature's deepest secrets, including unification. Although I disagree with Wilczek that beauty in Nature leads to unification, there is no question that symmetry is and has been a guiding principle in physics.

a hundred years from now. There is no theorem proving that we now know all there is to know about how the elementary particles of matter interact. On the contrary, our current knowledge of what goes on in the world of the very small is awash with open questions and uncertainties. The best that we can strive for is to construct a unified theory of our *current* knowledge, accepting that it will most probably be revised in the future.

This kind of view is often perceived as defeatist: "What's the point then, if we can't reach an end to our pursuits?" I find this kind of finalist mind-set to be profoundly misguided. It's the same as the fellow who won't go fishing because he can't be sure of catching any fish. It is precisely the uncertainty, the not knowing whether we will be successful, that brings excitement to the sport. Every bite is a surprise, the unexpected greeting you with the physical shaking of the rod. We all want to catch fish; but not knowing whether we will makes all the difference. Success tastes good only if we know failure.

The fact that our knowledge of the world is incomplete should not be seen as an intellectual weakness or a defeat of our reasoning powers. Instead, it should be seen as liberating: the incompleteness of knowledge frees us to explore the ocean of the unknown without the burden of finding some kind of final truth. Every meaningful discovery leads to new questions. The expectation that we can grasp some kind of final truth transforms the scientific pursuit of knowledge into a religious pursuit, given that only in religion are such notions of finality acceptable. Even if there is such a thing as the ultimate nature of reality — a notion that mirrors the reality of the infinite, thinkable but not reachable — we can't embrace it in its totality, with or without science. Science is an ongoing construction, a narrative we devise to make sense of the natural world. To set as its loftier goal the unveiling of some kind of final truth takes away from its exploratory excitement, overburdening it with a sense of religious quest. The vastness of Nature's wilderness should not be limited to climbing a single peak, as if it were a pilgrimage to a fixed destination.

Consider, for the sake of argument, that such an end is achievable and that one day we will get to it. Then what? Science reaches an end? We stop asking questions about Nature? This, to me, is a much sadder and defeatist state of affairs than the certainty that there will always be something new to learn, that the hunt is on and will stay on as long as we continue to ask questions about the unknown. (And get the much-needed funding to build the tools to test our hypotheses.)

This explains why it's okay if I can't figure out what happened that day in Copacabana. Something happened, the almost simultaneous breaking of two separate containers of crystal glasses and bottles. I witnessed it with my very own eyes and was perfectly sane and stable. Moments before, I felt some kind of physical discomfort, the shivering I mentioned, and a strange urge to go to the dining room. Why? I don't know. I don't usually shiver or have urges to go places around the house. *Yo no creo en las brujas pero que las hay, las hay* . . . Whatever it was, it crawled onto my perceived reality from some forsaken region of the ocean of the unknown, only to retreat back into oblivion before being unmasked, an intruder whose nature remains as mysterious to me now as it was almost forty years ago.

Only the single-minded hedgehog rationalist, who deems what doesn't fit his worldview either spurious or trivial, would dismiss such numinous events as some sort of freakish coincidence. This sort of outright dismissal seems to me to be more a symptom of fear than of conviction, like the ostrich that would rather hide its head underground than face what's right in front of it. (Disclaimer: real-life ostriches *don't* bury their heads in sand when scared!) The alternative seems terrifying: embrace the mystery, accept that there are things beyond what reason can grasp, unexplainable things. Can a scientist do this, though? Can we embrace the mystery and still pursue a rational description of the natural world? Of course we can. And how much more wonderful life is when we thread along both paths, allowing mystery to inspire us and reason to illuminate our way.

## The Allure of the Mysterious

The truth is, I was a very mystical teenager, in awe of the mysterious. I still feel this awe, even though my youthful mysticism has now grown into a deep spiritual connection with Nature. In this I am in very good company. Let me quote Einstein: "The fairest thing we can experience is the mysterious. It is the fundamental emotion which stands at the cradle of all true art and all science. He who does not know it and can no longer wonder, no longer feel amazement, is as good as dead, a snuffed-out candle." And again: "I have never imputed to Nature a purpose or a goal, or anything that could be understood as anthropomorphic. What I see in Nature is a magnificent structure that we can comprehend only very imperfectly, and that must fill a thinking person with a feeling of humility. This is a genuinely religious feeling that has nothing to do with mysticism." My encounter with the witch of Copacabana didn't help boost the cause for a rational explanation of reality, even if by then physics was already my main academic interest. To a budding scientist, the incident had revelatory power: trust your reason, but not exclusively. The Universe is rational, but we can't ever grasp it in its entirety, being "a magnificent structure that we can comprehend only very imperfectly." There is the unknown, and there is the unknowable. I was very confused. Possibly, I chose physics as an attempt to find a legitimate basis for my attraction to the unknown, as a way to engage with the supernatural, to prove that it was really there, lurking under the surface of reality. This is not as crazy as it seems. Many distinguished scientists have tried to use science to decipher mysteries of the beyond, notably some of the greatest of Victorian physicists, men that include Lord Rayleigh, who explained the color of the sky, and J. J. Thomson, discoverer of the electron, as well as William Ramsay, Sir Oliver Lodge, and Sir William Crookes. Telepathy, psychokinesis (making objects move with the power of the mind, like the "force" in *Star Wars*), ghosts and communications with the dead, all were fair topics of investigation. By the late nineteenth century, a rising tide of spiritism had swept through Great Britain and the United States. Going to a séance was a popular night out. People who missed their loved ones entered the shadowy, candlelit rooms with their hearts filled with hope. Maybe they would get a sign, a message from beyond. Wouldn't it be wonderful if our short lives on this planet were only a small sliver of a timeless existence? If the dead were not truly dead but only departed from this realm? How could such a possibility not seduce us? If invisible electromagnetic vibrations filled space, what else could be lurking in this ethereal realm, parallel to the senses? Even if nowadays much of this sounds like humbug, our need to believe in the impalpable remains as solid as ever. Consider the New Age culture, excited about quantum healing, the power of crystals, auras, reincarnation, and so forth. Science remains the credible outlet, being used to support some very nonsensical ideas, often applied completely out of context. To my excitable seventeen-year-old mind, the mysterious was beckoning, and science was the safest portal to the unknown.

The witch of Copacabana so disturbed me that I grabbed on to the scientifically concrete as a shipwreck survivor to a floating log. If we live in a demon-haunted world, let's lock ourselves inside the church of reason, making sure, however, that we leave one small window unlocked. My choice was science, which I decided was the only way that I could turn the supernatural into the natural. At least in part.

For many years after this mysterious event, I wavered between atheism and agnosticism. Now, I am a professed agnostic. Atheism — even though probably correct in its core assumption — is too dogmatic in its absolute rejection of God: its foundation is a belief in non-belief, which I consider a contradiction. In fact, I go even further and state that atheism is inconsistent with the scientific method — no doubt a shocker to those who take it at face value. The nonexistence of God is not a question that can be settled empirically, a point recently emphasized by French philosopher André Comte-Sponville in *The Little Book of Atheist Spirituality*. "The absence of evidence is not evidence of absence" — a saying Carl Sagan made famous in relation to the existence of intelligent aliens. This is an essential point that strict atheists often overlook, as they conflate lack of evidence with their belief in non-belief as *proof* of nonexistence. Atheists who, at this point, are probably feeling quite enraged will hopefully take a deep breath and continue reading.

All my training in science, plus the sum total of my life experiences (thus far), leaves me with no other viable choice. I don't see any reason to believe in God or in the existence of the soul, but I cannot *absolutely* rule either out. I can hear some of my religious friends declaring me a fool: "Even after witnessing an out-of-this-world event, he clings to reason. How pathetic!" I can also hear some of my atheist friends saying I'm an even greater fool: "What sort of rational being would give any credence to these fairy tales that have ripped civilization apart for thousands of years?"

Let me clarify. Agnostic: someone who doesn't believe in any supernatural deity, but doesn't categorically deny it either. This is a softer position than *strong agnosticism*, which states that an agnostic is someone who believes that nothing is or can be known about the existence of God or about anything beyond material phenomena. This position, again, is too dogmatic in the light of science. To state that something can't ever be known about a topic is a very dangerous position to take, as the history of science itself has shown. What may look now as "beyond material phenomena" may turn out to have a perfectly materialistic explanation in the future. The boundaries of what we call material get enlarged all the time. We are currently surrounded by "dark materials," to use Philip Pullman's nomenclature, dark matter and dark energy, cosmic substances of unknown composition. What are these substances that control the cosmic dynamics? We don't know and, a few decades back, didn't even suspect their existence.*

Take, as another example, quantum entanglement, the property of certain quantum systems that Einstein used to call "spooky action-at-a-distance," as it seems to violate the notion that the speed of light is the limiting speed in Nature. Here we have pairs or groups of quantum objects (electrons, photons, other quantum particles), which, even if separated by huge distances, respond to one another's presence as if behaving as a single entity. Let me illustrate. Imagine twin brothers who have only a green and a red shirt. One lives in the United States, the other in Brazil. Every day

---

*The first inference that dark matter existed dates back to the observations of the Swiss American astronomer Fritz Zwicky in the 1930s. Zwicky mapped the motions of galaxies in galaxy clusters and noted that their velocities were much higher than would be expected if what pulled them around was just visible matter.

they get dressed at the same time to go to work. The odd thing is that they *always* wear different color shirts; if one chooses the red shirt, the other chooses the green shirt, and vice versa. Instantaneously, without communicating. That's what entangled particles do, across time and space. In a sense, entanglement denies the existence of space and time, as the same effects happen at one yard or at a thousand miles away, and apparently instantaneously. (At least within the precision of the measurements.) These are called nonlocal interactions, as they don't seem to require any local cause, defying our usual description of things happening because some cause or force makes them happen, as when we kick a ball to make it move. Although physicists use quantum entanglement all the time in a growing number of practical applications, from secure bank transfers to the first implementations of quantum computers, we remain in the dark as to what's going on, being as spooked as Einstein, even after all these years.

As Thomas Huxley, who coined the word "agnostic," wrote in 1860, "I neither affirm nor deny the immortality of man. I see no reason for believing it but, on the other hand, I have no means of disproving it. . . . Give me a scintilla of evidence, and I am ready to jump at [the immortality of man]." To my mind, this mild form of agnosticism, shared also by Bertrand Russell, is the only one rigorously in accord with the scientific method. In a strict sense, we know only what we can measure and observe. The rest is unproved conjecture. And since our measurements and observations of physical reality are necessarily limited, so is our knowledge of what's out there. We don't see atoms and electrons, but infer their existence from the clicks and bumps that we see in our detectors. There is a huge gap between their realm of existence and ours.

Furthermore, even our best measurements are interpreted

statistically, and must be analyzed with great care. For a concrete example, consider how modern-day particle physicists determine whether a particle "exists." We can't see a proton or a muon or a Higgs boson. Their existence is inferred through data we collect. Take the discovery of the Higgs boson, announced at the European Center for Nuclear Research (CERN) on July 4, 2012. Signals were collected in two huge particle detectors, which are essentially amazingly powerful microscopes. The signals amount to bumps in the collision data at a certain energy scale. In order to associate these bumps with a real particle, the likelihood that they could be explained by other, more mundane events must be roughly less than one in 3.5 million, what is known as a five-sigma result in statistics. (The "sigma" is the Greek letter denoting standard deviation, the deviation from an established average or mean in a data sample. In a perfect bell-shaped curve displaying the collected data, 68 percent of the data are within one sigma from the mean, 95 percent are within two sigma, and so on. So, the more sigmas away from the mean, the rarer the event is.) To announce that the Higgs exists, scientists had to take the statistical variability of their data very seriously, including possible errors with the detector hardware. After careful analysis, a large group of scientists — thousands of them in the case of the Higgs — come together to collectively determine that the statistical evidence is sufficient to announce the discovery of a new entity. Not surprisingly, such decisions are often contentious; unanimity is rare. Does this mean we shouldn't trust the scientists? We definitely should. Certainly more than fishermen, anyway.

The discerning factor with a scientific decision is that it is never final. With science there is no "case is closed." Given that science is an ongoing search for knowledge, the continuous accumulation of evidence, including further experiments and data analyses, is absolutely key to its functioning. In the long run, things tend to work out just fine, even if there are bumps along the way. The bumps, the revisions, are how new knowledge is forged. A scientific discovery is a decision based on collective consensus after enough time has passed. Case in point: the very day that I was writing these lines (November 10, 2014), news came out of a paper published in the technical journal *Physical Review D* (where most of my scientific papers are published) contending that the CERN data don't necessarily point to the Higgs boson as originally proposed in the 1960s. Rather, the authors claim the data are equally consistent with a related particle, a kind of composite Higgs made of smaller constituent particles called techni-quarks. "The current data is not precise enough to determine exactly what the particle is," declared physicist Mads Toudal Frandsen in a press release. "It could be a number of other known particles." So, something is there, but we aren't sure what it is yet. And this is two years after the Higgs discovery announcement. Only further research and more data can elucidate the issue.

If knowledge is light, it is surrounded by perennial darkness. Or, as the French philosopher Bernard le Bovier de Fontenelle wrote in the late seventeenth century, "Our philosophy [read "science"] is the result of two things only, curiosity and short-sightedness." We want to see more than we can. Much of human knowledge results from the creative tension between want and can't. Since we can't predict everything about the future, we can't know for sure all that could happen. A few steps into the ocean of the unknown and we are lost. Back to atheism versus agnosticism: if one day I see *hard evidence* of supernatural causes or ghostly beings in the world, or of the immortality of the soul, I'll be happy to side with Thomas Huxley and change my mind. (I imagine most lucid atheists would too.)

There is, however, an essential caveat with any supernatural "sighting," as we briefly mentioned before. Any real-world manifestation of otherworldly phenomena immediately makes it this-worldly—that is, part of our material reality. If you hear, see, or smell something, this something has had a direct physical exchange with your sensory organs. For this to happen, it must have emitted information that your brain or equipment captured, be it through light, sound, or touch vibrations. A ghost that is seen is a ghost that is very much part of our observable reality. This being the case, scientific reasoning should be able to describe its properties. In other words, a ghost that is seen is not a ghost. At least not a ghost that is defined as being part of some immaterial realm. There is a fundamental issue here, one that precludes any chances of establishing contact with this parallel supernatural realm. For if we could have contact, the supernatural would be perfectly natural, and thus very much part of our reality. Ergo, we can't *ever* have evidence of the supernatural, unless we abandon the notion that the supernatural belongs to a realm parallel and inaccessible to ours. But if we do let go of this notion, the supernatural becomes part of the natural, although it may lurk in the shadows of the known.

Believers would question my use above of the expression "hard evidence." What is "hard evidence"? I define it as evidence not based on subjective experience, hearsay, or suspicious visions induced by ritual trance or deep isolation. It is evidence that others can verify, evidence that, ideally at least, is repeatable. Under this prism, my witnessing of Maria's presumed curse, being an isolated event, doesn't qualify as hard evidence. If she could do it again, and I saw it happen, I'd concede that, yes, there are mysterious powers at work. (And would rush to study them.)

Unfortunately, these weird phenomena never seem to happen

twice. Why do ghosts disappear when scientific instruments are brought to the haunted house? Why are angels and demons seen only in private? Why has no dead person ever come back to tell about the afterlife in any believable way? For some many years, I had passionately wanted to believe. When I was that young fisher-boy, I would spend many a night awake, desperate to see my dead mother, waiting for a sign, however small, that she still knew of me, that she somehow still cared about me. She died when I was six, leaving an emotional void that can never be filled. I look at my own five children now, how attached they are to their mothers, and wonder how I survived the loss. I was the "poor boy with no mom" to my friends, the one whose life was forever marked by tragedy. Sometimes, my despair was so overwhelming that I made myself "see" her. There she was, pale and translucent, hovering at the end of the long corridor in our apartment, dressed in a flowing white gown, gesturing for me to approach her. But when I did, torn by fear and longing, she would vanish in thin air, like a rainbow made of hope.

There is rampant dishonesty in the world, too many crooks taking advantage of people's need to believe. Who wouldn't want to know that we live forever, that this life is just a snapshot of an eternal existence? Who wouldn't want to know whether heaven and hell are real and not just very elaborate inventions? I certainly would! But I want the evidence to be concrete, not something foolish or outlandish. As the great physicist Richard Feynman said, "I can live with doubt, and uncertainty, and not knowing. I think it's much more interesting to live not knowing than to have answers that might be wrong." The stakes are too high to allow myself to be taken in.

Even though I can't possibly bring myself to believe, I understand why so many do. Faced with the unknown, people typically have two choices: faith in the existence of the supernatural, or reliance on proof. I tried both and chose the latter. Most people choose to believe. In the United States, believers amount to about two-thirds of the population. To them, proof is more than unnecessary; it's a hope killer. You hear things like "You can't measure love, but it exists," or "You'd see that God is inside of you if only you'd allow yourself to feel him." If we call God love, sure, I feel his presence all the time. I feel him when I'm out in the river, water up to my waste, fly rod in hand. I feel him when I look at my children (especially when they are sleeping). But I don't see why we need to call love God. To me, calling love love works just as well. It's a real phenomenon, nothing supernatural about it. In fact, it's something the world could have more of.

And so, I came to realize that there is a third possibility: faced with some unknowns, and after trying to understand them with the methods of science, it's okay if we remain without an answer. I don't mean we should give up. This is not about defeat. After all, we will know only if we try. While some unknowns will be answered in the future, others surely won't, and new ones will certainly emerge. Even as the answers contribute a little to the Island of Knowledge, what should we do with those unknowns that after much trying and perseverance still remain unknowns? To me, the choice is clear. With Einstein and Feynman, we should embrace the mystery, embrace the not knowing, humbly accepting that our knowledge of the world, remarkable and ever-growing as it is, will always remain incomplete. This is not easy to do. Even Einstein, who clearly embraced the incompleteness of knowledge at an intellectual level, couldn't accept emotionally that quantum uncertainty was an unknown that couldn't be answered — that it was an unknowable aspect of reality.

We strive for light, always more light, but must understand that there will always be shadow. This choice — the dynamic complementarity between knowing and not knowing — brings me peace and feeds my search for meaning.

## A Line between Two Worlds

Back to Durham and fishing. Jeremy and I climbed the hills of the North Pennines for quite a distance. The roads got narrower, the fields became hillier and more barren, covered with thick, rough pastures. The sheep were still around, grazing peacefully within their stone-walled partitions, woolly pawns in an Escher chess game.

"There! You can see it behind those hills ahead of us, Cow Green Reservoir."

I thought I had left planet Earth. A very large lake, about two miles the long way across, with dark waters that could easily be hiding the monstrous Eachy, a cousin of the Creature of the Black Lagoon or one of its slimy friends, popular from local folktales. There wasn't a single tree in sight. The surrounding hills looked like the bent backs of sleeping giants holding up the gray, fast-moving sky. Rocks belted the shoreline, the boundary between our world and the trout world: invisible, cold, alien.

"Funny name, 'Cow Green,'" said Jeremy. "There are no cows, and nothing here is quite green."

"True," I nodded, a bit overwhelmed. The wind was picking up. After heavy downpours, strong winds are the next worse thing for fly-fishermen.

"This wind will make things a bit more challenging," said Jeremy. "But not to worry, I will show you how to cast against the wind."

Great, I thought. My first guided fly-fishing trip, and I have a challenge on my hands. I tried to psych myself. Embrace the challenge. It's how we learn.

Jeremy gave me waders with boots attached, a nine-foot, 5-weight rod, and tied two flies to the leader with amazing dexterity, the smaller one as a dropper.*

"Droppers are good, you have twice the chance of hitting on the right flies, you know, the ones the fish like. We will use these black-headed nymphs. They always work here." I marveled at how the trout could actually see such little black things from under the reservoir's dark waters. I wished I could see mosquitoes in a dark room.

I walked toward the lake and waded in, feeling a bit sheepish about the whole thing. There was no one in sight. Dark wave crests crisscrossed the water surface. It was getting windier and colder by the second. I wondered how long the giants would manage to keep holding up the sky before it came pouring down on us.

"Not too far in. They are all close to the edge."

I cast for the first time. Disastrous. I cast again. Pretty bad. The rod feels too light, the wind carries the flies sideways, and the dropper further complicates everything, as the hooks can get tangled with each other and to the leader.

"Marcelo, beat the rod against the wind. Hold it with your thumb up, not with your index finger. Accelerate the rod on the down cast!" The thumb points up. Of course! Why on earth did no one ever tell me this before?

My casting gets better. I cast a few times. No fish. "Teeming with trout?" I mumble to myself.

"Okay, let's move toward that rocky outcrop to the left."

---

*A dropper is a fly tied to another one, usually at the curving of the hook.

We do. I struggle with my cast still, but I manage. It begins to feel more natural, wind and all.

Suddenly, I feel a hit, an electric shock along the rod. The trout is hooked. I panic. My fly line is all over the place. By the time I start retrieving, the trout is gone.

Jeremy smiles knowingly from a distance. "Marcelo, you have to manage your line. Always retrieve it quickly after you cast, always keep it tight. Any slack and the trout will shake itself off the hook."

I practice retrieving the line using my index finger as a guide. It works, sort of. At least now I know there are fish lurking in these dark waters. My heart is pumping hard. I realize how far I am from a state of grace, that elusive state of oneness where you lose yourself in the water. Even so, for the first time I could sense that this state exists, that it is achievable. One day.

Few experiences in life can be as humbling as learning how to fly-fish. I know I mentioned this before, but it's worth returning to it again. It seems nearly impossible to keep everything in check. There's the choice of fly, the casting, the rod positioning, the line tightness; then there are many variables to watch for: wind direction and strength, water motion and depth, obstacles like rocks and trees. You are trying to penetrate into an alien world without actually going there. All you have is a line connecting two worlds, the bridge between our oxygen-saturated reality and the watery realm below. You must foresee what the fish will do, be completely attuned to their moves, guess where they are hiding, what they like to eat. You must turn into a being that exists in both worlds at once, not an easy thing to do. But exhilarating.

I look around. The stark beauty of the place is awe-inspiring. I couldn't have asked for a more extraordinary spot. As I am lost in contemplation, not something you should do for long when you fish, I feel a trickle running down my legs. No, I'm not peeing in my pants with excitement. It's a tear. A tiny tear in my waders. Damn! Within seconds, my feet are completely drenched. I start to shiver.

"Don't cast against the wind. Cast diagonally," instructed Jeremy.

And so I did. Almost immediately, another jolt. But this time, my line was tight between my fingers. I retrieved it slowly, as the pretty brownie leapt two feet into the air. Yes! I caught my first trout at Lake District. She wasn't too big, about twenty-two centimeters, a bit over eight inches. But gorgeous, golden brown, black and red spots along her sleek body, strong and healthy looking. The boy smiled approvingly. I hastened to unhook and release the fish.

"Wait!" called Jeremy "A picture. We must register this! You got your first brownie!" My mentor smiled with visible relief, camera in hand. "Okay, keep wading down, never cast on the same spot. Brownies never give you a second chance."

I finished the day catching two, losing four, happy, cold, and very wet. I learned a huge amount. About how to fly-fish, about endurance, and about how much I had to learn. Precisely what I needed.

## A Path with a Heart

I thrive on being challenged. Delayed gratification, nothing like it. Every fisherman knows it; every athlete knows it; every scientist knows it. It's a pervasive state of mind that colors everything you do. When I was twelve, I started playing volleyball in Rio. There was nothing I wanted more than to be a good player, to be part of one of the local teams fighting for the city's junior championship. It was a struggle, as I wasn't particularly intimate with the ball. Other players teased me to no end. I was bullied on many occasions, and even got the nickname "spider-duck" because of my awkward way of waving my hand as I was spiking (hence the "duck") and the many times I ended up tangled in the net (hence the "spider"). Fortunately, having two older brothers, I was no stranger to verbal abuse. I trained hard six days a week, including Tuesdays and Thursdays with the girls, a routine that brought me many added benefits.

It took me about two years, many humiliating episodes, but I learned. I was never the best of the team, but I didn't care. I was *on* the team, playing, winning, losing, being yelled at by my coach, a stiff army major who, I now realize, taught us how to focus relentlessly on an end goal. To prepare us for the championship, he made us play for a whole year in an older age division, so that we would face guys much bigger and stronger than we were. We lost almost all our games, sometimes by humiliating scores. We were pounded. But when the time came for us to play in our age group, we were a tightly knit team, hungry for victory. We won every game, two years in a row. I even went to play for the state of Rio de Janeiro in the Brazilian national championship. Our setter was none other than Bernardo Rezende, the famous Bernardinho, who was to become volleyball's most successful coach in world history. We won the nationals, beating São Paulo by three sets to two in the final after being behind by two sets. The spider-duck became Brazil's junior champ.

This is a lesson I never forgot.

The same happened with my decision to study physics. How does someone decide to become a physicist? Why did *I* become a physicist? What does it even mean, to *be* a physicist? Everyone understands what it means to be a doctor, a lawyer, an engineer, or a stockbroker. Even a chemist or a biologist, as they can work for the pharmaceutical industry, at various industrial applications, practical things. What does a physicist actually do?

The essential task of a physicist is to uncover the fundamental laws of Nature. We do this by investigating the behavior and properties of all physical systems, from subatomic particles and different materials to fluids, stars, and the Universe as a whole. Most of us teach, as lecturers or mentoring graduate students; the majority work at applied research, in the computer and aerospace industry, developing new technologies and materials, or in consulting and finance, developing mathematical models for risk management and hedge funds. (Several of my PhD students have gone this way; I'm counting on them to fund my research group when they become multimillionaires.) Some work for national labs making bombs, others for different areas of the defense sector. There are many applied jobs for physicists, some blurring the lines between physics and engineering. But as a teenager, these were *not* the kinds of physics jobs I had in mind. I was thinking of Einstein, Bohr, Newton — the pioneers, the visionary geniuses that essentially defined the way we think about the world and, through the applied spin-offs of their basic research, how we live our lives. *That* was the kind of physics that impressed me: theory, the fundamental questions, the science that unveiled Nature's hidden secrets, that engaged with the mystery of existence.

My father was quick to burst my bubble.

"Are you insane? Do you even know a real physicist?" he yelled when, at seventeen, I solemnly announced I wanted to get a physics degree. "And who is going to pay you to count stars?"

"But Dad, physicists don't actually count st — "

"We live in Brazil, not in England or the US! Brazil needs engineers! Get a real job! Go to engineering school."

And so I did. Chemical engineering, to be precise, at the Federal University of Rio de Janeiro. It didn't take long for me to find out that it wasn't going to work out. My performance in the chemistry lab was abysmal. I would have failed the class if the theory portion didn't carry half of the final grade. Calculus and physics, on the other hand, were a breeze. And fun. I knew I had to do something, with or without my father's approval. Early in my second year, I got a small grant from the Brazilian government to study the theory of relativity with a physics professor.* Within a couple of weeks, I knew there was no turning back. I completed the second year of engineering and quit, transferring to the physics program at the Pontific Catholic University, at the time the top physics department in the country. It was the scariest — and best — decision of my life. I was alone at the beach again, facing an unknown ocean, beckoning me to step in.

Was I "good enough"? What were my goals? My father's words kept echoing in my head. What if he was right? *Who* was going to pay me to count stars? At the time, being a physicist in Brazil meant something quite different from being one in the United States or Europe. Either you got one of the few university jobs available, or you were unemployed. The competition was fierce.

The way I saw it, I had no choice. I mean, of course I had a choice, at least in principle. But I knew that this was it, that this was my path, it didn't matter what it took or how hard it was going

---

*This is an amazing federal program that is still going on, providing small grants for "scientific initiation." The student finds an adviser, and together they decide on a project to work on for a year. It's a one-to-one apprenticeship that brings a young student in direct contact with the world of learning and mentorship. It changed my life and, I'm sure, that of thousands of young Brazilians as well.

to be. It didn't matter that I was no Einstein or Bohr. What mattered was that I was following my heart. Only after I had made up my mind did I feel whole again. If choices in life are based on how easy and safe they are, life will be easy and safe: a boring, half-dead existence. I saw how my father and his friends complained bitterly about their work, how unhappy they were trying to fit into a life that their parents and circumstances had predesigned for them. Was that what growing up was about? I wanted the passion, the adventure, the uncertainty. You go for the big fish not knowing if you'll ever catch one. Every cast holds a new promise. It is true that most of the time you catch nothing; you make mistakes, the fish escapes. But if you are persistent, if you keep the fire burning in your belly, sooner or later you'll see the rewards. Not necessarily by catching a big fish, but by the fishing itself. We grow by doing. We live by doing. With every cast, the line goes farther and you get closer to yourself. That was a key discovery for me, to live a life of meaning.

# 2

# São José dos Ausentes,
# Rio Grande do Sul, Brazil

## Tropical Trout

Since Brazil is my native country, I go back a lot. Often not as a
research scientist, although occasionally I do give technical talks
at universities and conferences. I go back mostly to lecture about
science to the general public. I've had my own weekly newspa-
per column for nineteen years and have presented two TV series
for more than thirty million viewers. I travel the country talking

about the Big Bang and black holes and the Higgs particle, about the relationship between science and culture, and between science and religion. I do this in the United States and elsewhere, but in Brazil I am one of the few scientists actively engaged with the public understanding of science. I wish there were more of us. I do this for many reasons, but most importantly because I firmly believe that science belongs to society, that scientists generate culture, and that this culture should be shared and discussed openly with everyone. I do it because science shapes how we live in essential ways and is intimately enmeshed with our future: if we close our eyes to the major science-related issues of today — energy resources, global warming, the water supply, the dos and don'ts of genetic engineering, the spread of nuclear weapons — we will pay a very high price later. Even worse, our children will be the ones paying for our bad decisions.

Science education in Brazil, and to a lesser (but still very problematic) extent in the United States, is in bad shape.* In Brazil, seven out of ten public school physics teachers neither are physicists nor have a degree in the physical sciences. Most are biologists, geographers, or language teachers. Many don't even like physics. How can a teacher who is not professionally trained and dislikes his subject instill a passion for learning in young minds? It's a true tragedy. Countries without well-trained scientists and engineers are fated to become technologically dependent, to lag behind in this very competitive digital age. At the individual level, anyone deprived of a scientific view of Nature misses out on one of the most magnificent of human achievements, as passionate and life-

---

*A recent survey from the National Science Foundation found that about a quarter of Americans didn't know if the Earth moved around the Sun or vice versa. About one-third of Americans deny the theory of evolution or that we evolved from apes.

transforming as art, literature, and music. Shakespeare, Impressionism, and Einstein should be part of every school's curriculum.

The shame of it is that children love science. They are born scientists, always mixing things and throwing things, experimenting, often to the exasperation of their caretakers. This finding-out-about-the-world tends to last until they are about twelve or so, when the onslaught of hormones drives them away from "why" questions, spurring instead an all-consuming interest in sex. Pheromones are a tough opponent for any educator.

What is needed is a change of approach. Science is about Nature. It's about solving problems by learning how Nature solves problems, how it operates. As such, it can't be taught solely in classrooms, on blackboards, or in virtual computer-based experiments. Children must *see* Nature in action in order to fall in love with it and to want to understand it. They must be taken outside to observe the staggering variety of motions, shapes, and transformations that take place at all scales, from the microscopic to the cosmic. They must experience the remarkable creativity of life, the way energy flows from the Sun to the atmosphere to the oceans to the plants and to us. They must witness the interdependence of biological, chemical, and physical systems, how they work in combination, oblivious to our artificial compartmentalization of knowledge. Before students learn science inside the classroom, they should first experience Nature at close hand: observe, engage, and only then conceptualize.

A few years back, I went on a very special trip to Brazil. On the surface, it was a typical public lecture tour, hopping from city to city. But this time I added something different. The trip was near its end, after two weeks of a grueling but rewarding schedule. My last stop was Porto Alegre, the capital of Brazil's southernmost state, Rio Grande do Sul, home of outstanding *churrasco*—the

Brazilian-style meat barbecue — and *chimarrão*, a bitter tea that warms the gauchos in the cold winter nights of the pampas.

To entice me to go all the way to Porto Alegre, one of the hosts told me about trout fishing there. (I guess my love for fishing isn't a secret in Brazil.) Trout fishing? In the Brazilian tropics? I was skeptical, but my host confirmed it. Not just trout fishing, but in streams designated fly-fishing only! I had never heard of fly-fishing in Brazil. The few Brazilians who know what fly-fishing is had seen it in *A River Runs through It* or some other movie. (I am guilty as charged, as the movie was my first contact with the sport. There was raw poetry in the way Brad Pitt's character engaged with the flowing water, with Nature, with the prize fish.)*

Ever the skeptic, I consulted the oracle (Google) about "fly-fishing in Brazil." To my joyous surprise, my host was right. Tucked in the middle of a little-known mountain range in the northern part of the state are rivers filled with rainbow trout, and some are indeed for catch-and-release fly-fishing only. Unbelievable. After many phone calls and e-mails, it was all arranged. Alexandre, my guide, would pick me up at 10 PM, right after my lecture. We would drive straight to a local inn, Pousada Potreirinhos, hoping to arrive by 3 AM to catch some sleep before taking off to the river in the early morning. At least that was the plan. It all depended on the state of the very nasty dirt roads, which often flooded badly that time of the year. Assuming we made it to the inn, it was up at 6 AM for a quick home-cooked breakfast so that we could hit the river by seven. I don't think I had ever looked so often at my watch while lecturing as I did that night.

---

*There are now numerous fly-fishing excursions to the Amazon River basin, especially to catch the amazing peacock bass — a very different experience from trout fishing.

## Changing Your Worldview Is Hard to Do

I talked about *The Harmony of the World*, my first (and so far only) novel, which had recently been published in Portuguese to become a surprising best seller, a fictionalized retelling of the life and work of the brilliant seventeenth-century German astronomer Johannes Kepler, one of my all-time heroes. (If you're going to spend three years of your life writing about someone, it had better be someone whose life you believe is worth retelling.)

There were about two hundred people in the audience, from all walks of life. My presentation was part of Porto Alegre's popular annual book fair, an open-air celebration of reading, with dozens of daily lectures and book signings. The theme was not my novel per se but an investigation of our changing views of the cosmos, from prescientific cultures to modern times. The emphasis was on *changing views*. The cosmos that we live in today is very different from that of the fifteenth or sixteenth centuries. Not the cosmos itself, of course, as it is what it is, but the way we think about it. For Columbus, as for everyone in Europe around 1500, the unmoving Earth was the center of creation. Concentric spheres, made of crystal, surrounded Earth like the layers of an onion, carrying the Moon, the Sun, and the five known planets in their celestial orbits: Mercury, Venus, Earth, Mars, Jupiter, and Saturn. The outermost layer was the sphere of the fixed stars, sealing up the finite cosmos. Outside sat God, the unmoved mover who gave rise to all motions. Deep underground, and thus much closer to us, the devil tortured sinners and lost souls. For Columbus and his contemporaries, hell was the true center of the cosmos.

To a large extent, this cosmic structure determined how people lived, as it blended the celestial arrangement with the vertical hierarchy of medieval Christian theology. The way of the cosmos mirrored the pilgrimage of the human soul as it strove to rejoin God and the chosen in heavenly afterlife. Medieval cathedrals, with their ascending architecture, beautifully illustrate this covenant, compelling the faithful to look up in awe, toward salvation high above.

Everything made sense. At the center of the cosmos and made of the four terrestrial elements—water, earth, air, and fire—our world was different from all the other worlds in the heavens, which themselves were made of the fifth essence or "quintessence," the eternal, unchanging ether. In this vertically ordered cosmos, the moralistic and the geographic commingled, creating a directive everyone understood and feared.

The slow collapse began in 1543, when a shy Polish deacon, Nicolaus Copernicus, proposed an alternative view, in which the Sun was the center of everything. Earth was pushed aside, turned into just another celestial wanderer, a planet of no great consequence. This rearrangement tore apart two thousand years of an Earth-centered cosmos, creating more questions than answers. If Earth was not the center of creation, were we? What rules dictated the order of things? Why was Earth the third planet from the Sun? Would hell be in the Sun now? Were we still God's special creation if we lived onboard a mere planet circling the Sun? I discussed how a shift in our cosmic view is a dramatic life-transforming experience, leading to a reorganization of beliefs, to a rethinking of our position in the grand scheme of things, to a redefinition of our role and mission in life.

The Copernican cosmos also raised more practical questions: if the Earth turns upon itself once a day, we are all spinning at about 1,070 miles per hour. Why don't we feel this? Why don't we get tossed off into space, as we would on a crazy-fast merry-go-round? Why aren't clouds and birds left behind? Copernicus had no firm answers to these questions. There was very little physics in his groundbreaking book. Planets were ordered based on the time they took to complete one revolution around the Sun: Mercury (three months); Venus (eight months); Earth (one year); Mars (two years); Jupiter (twelve years); Saturn (twenty-nine years). It made sense, and it fitted well with the aesthetics of the Renaissance, where order, symmetry, and proportion were equated with beauty. Copernicus attempted to add theological significance to the new cosmic arrangement, deeming it the inevitable creation of a God whose elegant aesthetics ought to be stamped in the cosmic blueprint: the cosmic aesthetics mirrored God's perfect mind.

A long time passed, however, until the new worldview began to take hold. Everyone who has gone through a divorce knows it doesn't happen overnight. There's too much pain and uncertainty. When you announce your intentions to break up with the past, very few people back you up. You feel very lonely. As the historian of astronomy Owen Gingerich noted in *The Book Nobody Read*, fifty years after the publication of Copernicus's book, only about ten people had expressed positive opinions about a Sun-centered cosmos. Chief among them were Kepler and Galileo Galilei, working in tandem during the first decades of the seventeenth century. The new worldview became inevitable only in 1686, with the publication of *Principia*, where Isaac Newton enunciated in clear mathematical prose the laws describing motion on Earth and in the skies. He went beyond that, describing how gravity ruled the mechanics of the cosmos and the falling of earthly objects, simultaneously. The same physics united the terrestrial and the celestial.*

*It is worth noting that the terrestrial and the celestial were united in two disciplines that predated modern science: astrology — where cosmic influences determined earthly matters at the individual level — and alchemy, with its motto "That which is below is like that which is above." Newton was no stranger to either,

Newtonian physics is the physics of everyday life, describing with great precision the motions we're all familiar with. Although some more complex phenomena, such as the soliton John Scott Russell saw rushing along a Scottish channel, needed more sophisticated treatments, they were all spin-offs of Newtonian mechanics. For example, Newtonian mechanics explains how properly applying physical principles can vastly improve your fly-casting ability. What casting instructors call "loading the rod" simply means bending it back as much as possible so that it can propel the fly line ahead more efficiently. This movement transfers the elastic potential energy stored in the bent rod into the kinetic energy (the energy of motion) of the fly line, as with a loaded spring. The more efficient the conversion of energy, the more powerful the cast and the farther the line goes. As Heraclitus put it some twenty-five centuries ago, "The bow must bend back to propel the arrow forward."

While Isaac Newton was pondering the physical laws that describe how Nature works, another Izaak, by the family name of Walton, was perfecting the art of angling along English rivers. A biographer of John Donne and other British luminaries, late in life Walton retired to the country to fish and write. My enlightened guide Jeremy would surely agree with the wisdom of this decision. *The Compleat Angler,* Walton's famous fishing manual, came out in 1653 and is still in print. I wonder if the two great "Isaacs," the natural philosopher and the angler, ever crossed paths. Given Newton's asocial personality, I doubt it, although I imagine Walton must have caught quite a few trout along the river Cam.

---

being, in fact, a devout alchemist. He must have delighted in his discovery of universal gravitation and its union of earthly and celestial physics.

## Love in a Time of Science

In time, our understanding of the cosmos changed. But not most people's view of it. Natural philosophers were pushing for a new worldview, but people resisted. The new cosmic order did instill awe at the vastness of creation. But it also instilled fear — the fear of being alone in a godless Universe. Few expressed this as eloquently as the seventeenth-century French philosopher and mathematician Blaise Pascal, whose words predated Newton by decades: "When I consider the short duration of my life, swallowed up in the eternity before and after the little space that I fill, and even can see, engulfed in the infinite immensity of spaces of which I am ignorant, and which know me not, I am frightened, and am astonished at being here rather than there, why now rather than then. Who has put me here? By whose order and direction have this place and time been allotted to me?" The more science advanced, the less necessary an all-powerful God became. "I have no need for this hypothesis," uttered the astronomer and mathematician Simon de Laplace to Napoleon, reacting to the emperor's surprise at not finding any reference to the Creator in Laplace's *Celestial Mechanics*, published in five volumes between 1799 and 1825. The cosmos became a precise machine, clockwork unwinding according to strict mathematical laws. God's role was relegated to that of the clock maker, the maker of laws: once God created the world, it would follow its deterministic course without any further need for divine intervention. Needless to say, believers didn't take this very lightly. Could science really go this far in explaining the world, leaving out only a single mystery, the mystery of creation? A God whose only job was to create the world was a distant God, impersonal and uncaring. And what of free will? If the laws of Nature predetermined the unfolding of all events, no action or choice was truly free: the time you were born, which person you'd marry, your profession, your trials and tribulations . . . everything would be already written in the book of time. We would be mere automatons, blindly believing in our autonomy as individuals, when in fact our freedom was a mere illusion. In a clockwork cosmos we would be no more than puppets in a drama played on the cosmic stage, director unknown.

To make things worse, as if on cue, came Darwin's heavy blow. We are just evolved apes with shrunken tails, the evidence led him to conclude, not quite the semblance one would expect from creatures made in God's image. Many felt — and still feel — that science had robbed them of God and offered nothing but cold materialism in return. How could we sate our spiritual hunger in a clockwork cosmos, where everything seemed reducible to cold mathematical logic and precise physical laws? Where did love fit in this picture? You can't really blame the Romantics for revolting against this excessive rationalism.

This is the rift that has never been healed, the great spiritual void of our scientific age. What are people to do? How can people cope with this quandary?

Some ignore the scientific message altogether and surrender their rationality to some blind form of religious extremism, generally through complete submission to dogmatic orthodoxy. This we see in violent fundamentalist movements, such as Al Qaeda or ISIS, where killing those who oppose your belief system is morally justified in the name of faith, and in nonviolent but still fundamentalist groups such as ultraconservative evangelical Christians and Hasidic Jews, with belief systems firmly anchored in the distant past. I find it quite ironic to see an Orthodox rabbi, all decked in black as his peers have been for centuries, happily using a GPS, talking on a cell phone, or, when illness comes, taking antibiotics or going for radiation therapy. How is it that the technological offspring of quantum and relativistic physics may be conveniently used as needed but not the revolutionary worldview they brought forth? The same science used to build these gadgets is used to date fossils, Earth's age, and life's evolutionary trajectory from bacteria to people. It's mind-boggling. And yet, this eyes-tightly-shut perspective is the only option for an alarmingly large number of people, not just religious extremists.

Others consider that science serves to illuminate and even boost their faith, as it allows for a deeper understanding of God's work. Contrary to what many think, a large number of scientists worldwide are deeply religious, or religious to some degree, and find no conflict whatsoever between their faith and their science. They would claim, correctly, that some questions belong to science and others don't. (We will address later what some of these questions may be.) This is a tradition going back to the patriarchs of modern science, including Copernicus, Galileo, and, especially, Descartes, Kepler, and Newton. To them, science was a form of worship, a way to approach God's mind. We can trace this connection to much earlier, to Ptolemy and even Plato, who posed as the philosopher's main task the contemplation of eternal truths that God set forth in creation. And given how the pre-Socratic philosophers Parmenides and Pythagoras influenced Plato's thought, the relation between Nature's hidden code and God's mind dates back to the very beginning of Western philosophy.

Still others are not religious in a traditional sense but use pseudoscientific concepts to justify their attraction to ancient mysticism. This grants their belief system the feel of scientific credibility, which is good enough, unfortunately, to fool many people. I'm referring to some in the New Age movement who,

despite all their good intentions — universal love, mutual understanding, healing, connection, respect for differences — ground
their beliefs in a science pulled completely out of context. We
hear of quantum healing, aura therapy, Kirlian photography, ra-
dionics, therapeutic touch, and the like, practices using concepts
like "energy," "quantum," or "field" in ways that have very little to
do with their physics counterparts.* A lot of quackery takes advantage of people's emotional and physical needs. What a science
purist can do, apart from unmasking the charlatans (as the Amazing Randy does so well), is to insist on the correct application of
scientific concepts, trying to rescue them from misuse and abuse.

When a New Age practitioner talks about summoning the
"subtle energy" of the Universe, or enhancing your "vital energy"
or manipulating your "biofield," it should be clear that these are
not physical or scientific concepts but metaphorical notions created to describe a certain state of interaction between a person
and the environment or between a person and a practitioner.

Science already tells us that we are cosmic creations, animated
stardust capable of wondering about its origins, the way the Universe thinks about itself. Isn't that magical enough? Breathe in the
*prana*, feel the chi coursing through your chakras, find ways to expand your experience of being alive. That's what we should all be
after, trying to expand our mindfulness, to connect with a grander
reality, mysterious and unknowable. In this, scientists and mystics
are one, even if the ways they search for a connection with what is
beyond the sphere of the known are so profoundly different. Just
don't equate the coincidences of everyday life — meeting a long-
lost friend, saying the same thing as your partner, having a pre-

---

*The interested reader may enjoy reading the entry for "Energy" in *The Skeptic's
Dictionary*, http://www.skepdic.com/energy.html.

monition — with quantum entanglement or with a synchronicity due to a nonlocal energy field in the cosmos that *meant* these things to happen. Just experiencing the emotion of the event for what it is should be plenty, without the need for some pseudo-scientific justification based on some invisible governing power. Free yourself from the need for a top-down causation, for an explanatory principle for all that happens. Celebrate the simple beauty of the unexpected! The hard-earned credibility of science, the work of thousands of dedicated men and women around the globe over four centuries, should not be misused to seduce those who seek for a safe harbor.

Finally, the last group includes atheists and agnostics, people who find no need to add supernatural deities or events to their life experiences. I have written enough about them and their differences, including why I have reservations about the extreme atheist position, which I find antithetical to science. The point I want to make here is that atheists and agnostics are *not* antispiritual. This is essential to our discussion and to the widespread perception of the nonbeliever. The magic of science lies not in justifying supernatural connections between people and the cosmos; it lies in making the unknown knowable. It's in bringing us closer to Nature, in showing how the atoms that make up our bodies were forged in stars long before the Sun and the Earth existed, in showing that we are cosmic creations in dire need of preserving our planet and the rare and beautiful life that blankets it. The source of science's true spirituality lies in revealing the material connectedness between us and the Universe.

Instead, the general perception is that if you are an atheist or an agnostic you can't be a spiritual person. Wrong! Recall the words of Einstein quoted earlier, that the attraction to the mysterious, to what is beyond our grasp, is deeply spiritual. We humans have an essential urge to expand our boundaries, be they physical or metaphysical. We go about exploring the world and, in our age, outer space, with an insatiable wanderlust. This lust for more is not relegated to the physical plane. It lives in our minds too, as we feel the need to expand our perception of the real, to flirt with the impossible, to push knowledge beyond its current boundaries.

We can't bear to be imprisoned in the same small place, be it a physical or a mental cage. Think of a goldfish in a bowl. The poor creature not only is doomed to spend its life confined to a small space, but is also aware — insofar as a goldfish is aware of anything — of a reality out there, of which it can capture some dim contours. Beyond the glass is freedom, tantalizingly close, yet unreachable. But the world out there is also riskier, unknown, possibly deadly. Same with us humans, imprisoned in our planet and in our current state of knowledge. A jump into space can expand our spatial boundaries, but it can also be deadly. A jump into the realm of the unknown can expand the Island of Knowledge, but it will also generate more unknowns, possibly even a few unknowables. But jump we must, for the alternative is unacceptable. Who wants to spend life swimming around in circles in a small bowl?

Love, then, emerges triumphant from this view, as the force behind our urge to find meaning. Science does not preclude love; it actually needs it as its seed, the nucleating force of our personal and collective growth. This has nothing to do with oversentimentalizing science. What Einstein called our attraction to the mysterious, and I call our attraction to the unknown, is just another way of expressing love. For what could be more mysterious than the attraction you feel for someone you love, the conviction that life without this person would be as incomplete as being stuck alone in a small bowl? The "person" here can be another human or group of humans, or it could be Nature. The opposite of love is not hatred; it's oblivion.

It may even be possible to trace the feeling of loving someone or a group to some evolutionary survival strategy, to some selective advantage of altruism. But no rational explanation of love does it justice, nor could it. Like a joke that loses its oomph when the punch line is explained, love's power lies not in knowing where in the brain neurons are flashing or what hormones are flooding your bloodstream when you meet that "person" or experience a deeply moving connection; love's power lies in its being felt, in its being shared. No scientific explanation of a feeling, although important in its own right and certainly an essential area of research, will ever replace the actual subjective experience of it. The two are very different things. Eventually, it will be possible to induce specific feelings through neuronal and chemical intervention. To a certain extent, mood-altering drugs and medications, and direct electric stimulation of the cortex, already accomplish some of this. But knowing how to induce a feeling and *feeling the feeling* are two very different things. For the mystery will always be there: how *I* feel love and how *you* feel love, each experience unique and wonderfully unquantifiable.

## Freedom in Attachment

My lecture over, I rushed back to the hotel to meet my guide, Alexandre Bertolucci (perhaps a relation to the great Italian movie director?). Alexandre's parents immigrated from Italy early in the twentieth century, part of a huge exodus of Europeans — especially Italians and Germans — to the south of Brazil. The mixing between the immigrants and the locals, whites of Portuguese descent, blacks from Africa, native Brazilians, makes for a high density of absolutely gorgeous people. Even those who aren't the result of much mixing, like the supermodel Gisele Bündchen — who's pretty much all German — somehow benefit from the area.

A burly man in his early forties, Alexandre had a handshake of someone who'd much rather be wading the pristine waters of the Silveira River than sitting behind a desk. A fly-fishing enthusiast is definitely an oddity in Brazilian culture. I stared at him as one stares at a bizarre creature in a sixteenth-century cabinet of curiosities. How on earth did this fellow become a master fly-fisherman around here?

We left Porto Alegre around 10 PM. Looking up, I could see a few brave stars challenging the city's glare. Excellent! São José dos Ausentes is notoriously cold — some say the coldest spot in Brazil, nestled at about thirty-eight hundred feet in the mountains on the northern parts of Rio Grande do Sul. Although temperatures don't compare to our northern New England winters, to wade in rushing water when the air temperature is about forty Fahrenheit is no picnic anywhere. And that's what these pioneers do around here. The trout demand it, since their peak feeding activity happens in water temperatures ranging between fifty and sixty-eight degrees Fahrenheit. Of course, trout being trout, this is just a general pointer, not an absolute rule. There are also small variations, depending on the species. Brook and cutthroat trout seem to be a bit more resilient at lower temperatures. But whatever the trout, their metabolism does slow down for lower temperatures and their appetite tends to wane. I always take a thermometer with me and check the water temperature before I start. Not that I'd give up if the water is too warm or too cold. But at least if I am completely unsuccessful I have something concrete other than myself to blame.

In the Silveira River, water temperature and flow are optimal between March and October, the season peaking around July, the middle of the Brazilian winter. Ironically, this is also when we fish in the northern hemisphere. The difference is that we are warm up here while they are cold down there. Lucky for me, this was late October, and the cold air was mostly gone. In fact, the problem was that it was already becoming too hot. Trout absolutely *hate* warm water. Not just hate; they die when the temperature climbs above eighty degrees Fahrenheit or so, as I have witnessed many times with a broken heart. Such beautiful creatures are not meant to float belly-up in a hot tub.

"Nobody has been fishing for the past two weeks," said Alexandre. "The trout will be begging to get caught."

"Yes!" I agreed enthusiastically. "Can't complain about that."

"But first we got to get there. Prepare for a long ride."

I soon realized that Alexandre wasn't joking. It was 3:30 AM by the time we arrived at Pousada Potreirinhos, a rustic inn strategically located about two hundred yards from the river. The road was impossibly treacherous, with some fifty miles of very rough gravel and dirt at the end. We even had to ford a fairly wide river, as torrential rains had carried away the bridge the week before.

"Thank God it didn't rain much today or yesterday. Otherwise we would have to take a major one-hour detour. I'm not kidding you; last time I tried crossing the river here, my car died on me," Alexandre said, with a twinkle in his eye. This was living. I nodded, relieved that the river was gracious enough to allow us to cross it. Last thing I wanted was to get stuck midcurrent at 1 AM in a Volkswagen Fox.

I had planned to take a nap on the way, rest up from my crazy day that had started with a flight at 5:30 AM from Salvador, Bahia, a state that might as well be a different universe from the cold and mountainous Brazilian south. Fat chance! My head would not slow down for a second, bubbling with anticipation. A full day at a trout-filled river? And in my home country? Fly-fishing in Portuguese; that's something I never thought I'd do.

I was up by 6:45 AM (if I had slept at all), ready to go. The inn was surprisingly full, but not with fishermen. Ecotourism is booming in the region, and a small bus had arrived with twenty people eager to hike the expansive hills in search of breathtaking views of deep valleys and waterfalls.

From my bed, I could see the sun blasting through a crack in the heavy wooden blinds. Wonderful! The smell of freshly brewed coffee — brewed the old-fashioned way, with a cloth filter — and of homemade muffins and bread was beckoning. Alexandre was also up, even more excited than I was, like a kid going to Disneyland.

"Every time I fish feels like the first time; things are never the same," he said, his black Sicilian eyes already in full alert. Alexandre had found his paradise, his mode of worship, and was eager to share his experience. I could sense he was going to be a great mentor, even if very different from Jeremy of Cumbria. The only thing the two men had in common was a passion for fly-fishing.

After stuffing ourselves, we got the gear ready. This time I had brought my own waders and boots, to avoid any mishaps. No more cold wetness for me, thanks. Also, fly-fishing is still a very eccentric activity in Brazil — there are no shops to buy equipment, at least none in 2007. There are a couple of fly tiers in São Paulo, but not much else. Everything is improvised, materials are used and reused, repaired and re-repaired. US- or UK-made boots are precious possessions. Orvis is some kind of deity from the north.

The sun was already blasting at 7:30 AM, the air temperature about eighty degrees. "It's going to be a scorcher," said Alexandre. "But not to worry; the water will still be quite cold." I looked

around. The Silveira River cut deep through a postcard valley —
bright green pastures on rolling hills, dark granite peaks in the
distance, a lushness quite distinct from the starkness of Upper
Cumbria. Tropical fly-fishing! Live the paradox.

We walked down to the river. Not very wide, textbook-perfect
with large boulders, lots of riffles and places where trout like to
hide, waiting for the current to bring down their food. Alexandre
had brought two rods, a light 2-weight and a still light 4-weight. I
claimed the 4-weight, which for me at the time was already quite
a challenge to handle. He fitted it with a beaded nymph and, on a
dropper, with something they called *pâncora* (pronounced pahn-
kora), a bright red woolly thing that mimics a small river mollusk
that the local rainbow trout relish. "*Pâncoras* never fail," Alexandre said.

The tying of the flies took some three minutes; very long because the poor guy didn't have one of those nice pre-made taper-
ing nylon strings called leaders we take for granted here. He had
to fashion one from scratch, tying together a sequence of lines of
increasingly smaller diameters, until he reached the thin 0.16 centimeters (about 0.063 inches) he used to tie the two size-12 flies.

"Cast over there," instructed Alexandre, pointing to the opposite edge, where the current reached what physicists call a no-slip
condition — practically stationary at the river's edge, right under
a huge tree. Easier said than done. First I had to get used to casting with a 4-weight rod, a big change from my 6-weight, or even
Jeremy's 5-weight. It felt like I was holding a tube of air. But after
two failed attempts, I hit the right spot. To my amazement, the
rod bowed almost immediately.

"Hook it!" shouted Alexandre.

I yanked the rod backward, holding on to the fly line. Was this
really happening? A trout on my first cast? Even my rough rod-

and-line management couldn't screw this one up. She was a small but beautiful rainbow, eager to eat her *pâncora* breakfast.

"Well done," said Alexandre. "But try being a bit smoother retrieving the fish next time. A larger one won't be this easy to handle."

Prophetic words. I took a couple of steps downstream and cast along the same general direction. After a few seconds, I felt the telltale jolt on the line. The rod bent in half. It was a big one. I stated to retrieve the line. The trout protested, having her fate usurped from her, and moved downstream. She leaped out, her silver body flashing against the bright sun. *This* was real fly-fishing.

Too excited, I started to lose my cool, dropping my rod and letting the line flap about. "Keep the rod vertical, high up," said Alexandre. I tried. But every time I attempted to retrieve some line, the trout moved the other way. The rod was bending like crazy. The line was too taut. In my inexperienced eagerness, I forgot that lines snap, especially thin ones. And so it was — a final pull and the trout was gone. A four pounder, maybe more.

"You can't fish like this," said Alexandre, who was as disappointed as I was. "You have to let her move about, give her space, play her, tire her out."

I learned the hard way one of the most important lessons in fly-fishing, something folks with heavy reels and thick lines rarely experience: if you want your fish, give it freedom. Don't go nuts retrieving the line, trying to bring it in too quickly. I thought of my first love, a beautiful fifteen-year-old girl named Anete, and how I totally spooked her with my sixteen-year-old overeager intensity. How often do we want someone so badly that we overdo it, putting so much pressure on the relationship that we end up pushing the person away? "If you love somebody, set them free," sings Sting. What a difficult balancing act, to want so badly and to have to let go. Lovers and parents know this only too well.

The experience made me think of fishing as a game of seduction, a give and take, a metaphor for love. You pull on the fish too hard and the fish will pull back, trying to get away. If you want the fish near, you first must let it roam free, or at least let it think it is free to roam. (But watch for that line slack! Too much freedom and *bam*, it will be gone, leaving only frustration in its wake.) Soon it will stop and give you a chance to retrieve some line, and then a little more, until you are right next to one another. (Does the hook hurt the fish? Apparently not.*) The lesson is, be gentle. Fishing is not a tug of war with the fish. If you tug on its jaw too hard it will react to the pressure and fight to regain its freedom. If you insist, and the fish is a fighter, it will twist its head this way and that until it breaks loose from the hook, or it will snap the line and leave you empty-handed. In fishing and relationships, seduction should be a give-and-take, not a one-sided imposition. The other is another, with her own sense of attachment and freedom.

In losing that trout I learned to respect it. It didn't escape to mock me. It escaped because I didn't know how to make it feel free. Only in the truest relationships is there freedom in attachment.

Alexandre fixed the line, fastening another *pâncora*. I cast again and, in moments, had another small trout. That was three

---

*On the issue of fish feeling pain, they apparently don't, at least according to current understanding. They lack the neurophysiological capacity for a conscious awareness of pain; what we attribute as reaction to pain is behavior based on human criteria and thus prone to misinterpretation. These are the conclusions of a recent broad investigation that looked at previous studies on the matter. The reference is J. D. Rose, R. Arlinghaus, S. J. Cooke, B. K. Diggles, W. Sawynok, E. D. Stevens, and C. D. L. Wynne, "Can Fish Really Feel Pain?" *Fish and Fisheries*, 2012, doi:10.1111/faf.12010.

for three, absolutely unheard of in my short life as a fly-fisherman. Still, I felt humbled by the experience. I had hooked three trout in fifteen minutes, but lost the trophy one because of my inexperience. I was learning, and that's what mattered.

## Limits Are Triggers

As we moved downriver, things changed. Trout fishing is truly mystifying. In one spot someone can't stop hauling fish after fish; move a few yards away and nothing happens. Compare the two spots and they look almost identical; same water temperature, same kinds of boulders, current flow, depth, and same bugs. But what may look the same to us may as well be a different universe for the fish. As every fly-fishing instructor tells us, reading the river, taking all these variables into account, is the most important thing a fisherman needs to learn. If you have only a couple of hours in the river, you don't want to waste them in a hopeless spot. Jeremy, my mentor from Cumbria, taught me the ruthlessness of trout: "Trout are unforgiving; don't bother casting in the same spot twice." Although it's okay to try a couple of times on the same spot or thereabouts, even if to test different flies, after a few unproductive casts moving on is definitely a good idea.

Alexandre decided to explore a faster-moving section of the river, right after a small waterfall. He cast once and, in seconds, had a trout. "Ha! Here they are!" he said beaming, as if he had caught his first fish ever. My turn came and . . . nothing. I cast again . . . nothing. "Ok, let's move down a bit," Alexandre suggested. I cast here and there, this way and that. No hits. Impatient, Alexandre tried. No hits. "Okay, let's go for a walk, try a different spot upstream."

I looked up. The sun was blasting unashamedly, as it often

does in tropical latitudes. We had been fishing for four hours. The water was getting warmer by the minute. If I were a trout, I'd have retreated to deeper pools. Alexandre took us to a spot where the river was a bit wider, with large boulders creating slow-moving deep and cooler pools, a perfect hideout for the wary trout. I had a go or two, and Alexandre changed the fly to a dark beady thing, similar to a stonefly nymph. "Cast upstream and let it drift awhile so that it's fairly deep when it gets to the pool." Bingo! An electric pulse shook the rod almost immediately after the fly passed the boulder. "Careful retrieving the line, make it taut, rod up!" yelled Alexandre. I brought it in, a beautiful creature, maybe a two-pounder, reflecting the sunshine on its silvery surface. "Quick! Let's put it right back! A trout is too precious to kill." How true! A trout *is* too precious to kill, especially in a fly-fishing-only river. (In any river, really.) The trout you put back in the water today will be someone else's joy tomorrow. Who knows, it maybe even be your own.

Man is the most perverse of hunters, always going for the trophy animal, the alpha male, the largest of the herd. That's not how most predators behave. They go for the runt, the one left behind, the wounded. The way we hunt is highly destructive; by killing the strongest we weaken the gene pool, robbing from the herd its selective advantage. This should be obvious from watching how most animals reproduce. We watch documentary upon documentary of males fighting for dominance before the champion is allowed to mount the female. Why so? It's all for gene selection. The winner of the combat is the strongest; as he reproduces, his genes will pass on to the next generation. If we kill the strongest we are compromising the survival of the group, weakening it, interfering in a damaging way with the operations of natural selection. Hunters and fishermen who go after the largest prey and kill it, taking out the trophy salmon or bonefish or the six- or eight-point buck, are compromising the future of the herd. Fishing or hunting for sport, as opposed to survival, should follow a catch-and-release approach for fish and use tranquilizer shots for animals. There's always the tale-telling photo next to the twenty-five-pounder salmon or eight-point buck to show off to friends and family.

We spent some time around the same waters, but nothing else happened; things were clearly calming down. I could sense Alexandre's frustration mounting, as he moved us here and there. After an hour of this, I suggested we take a lunch break. "We can try in the afternoon, see what happens then."

"Not a good idea," he replied. "After lunch the water will be even warmer. You won't catch a thing."

"No problem," I said. "I'm happy with what happened so far. Maybe you could help me improve my casting technique and teach me a couple of things about the river."

"That works," Alexandre replied, somewhat relieved. "Let's go and eat!"

We just put into practice Lao Tzu's famous words: "If you do not change direction, you may end up where you are heading." We were headed for a very frustrating few hours; much better to use the time for something constructive instead.

After a delicious lunch of rice, black beans, and BBQ steak (I still ate meat then), we headed out for a productive casting lesson. I learned more in a couple of hours than in all my previous casting experiences put together. Holding the rod close to the body, thumb pointing up, wrist strong, rhythmic movement from 11 to 1 o'clock, waiting for the rod to reach its lowest point behind me before beginning the forward thrust, all this to maximize its pendular swing, resulting in a more efficient transfer of elastic energy stored in the rod into the kinetic energy of the fly line. An eager apprentice and a talented mentor are a wonderfully efficient combination. By midafternoon, what before was a cumbersome motion became more of a flow, the fly line bursting out of the rod, the embodiment of power and grace, to land gently fifty yards away. "I got this," I said smiling. "No," replied Alexandre. "You got better at it. You never really *get* casting; you only improve on it."

That's a lesson we can use in most activities, from sports to work. We may set goals as useful guidelines and to organize a strategy to achieve them. But once we get there, we can always go for more. More speed, more precision, more results, more data, more delicious meals if you like cooking, better casting. The danger is in becoming blinded by this "more more" drive and losing our sense of self, the understanding that there is also contentment in having reached a goal or level of achievement. Recall my grandfather's dictum: "If you wear a hat bigger than your head, it will cover your eyes." So we must find the right fit and keep on challenging ourselves without losing sense of where we are. We must change hats, not their size. There is a dynamic, healthy balance between wanting more and enjoying what we have achieved. We must hit against our limits to challenge them. Limits are not insurmountable obstacles but triggers, the fuel that propels us ahead. Where we find ourselves depends on how far we want to get.

If we don't challenge ourselves we stagnate. But if all we do is to challenge ourselves, blindly trying to get farther than we are, we never enjoy what we have achieved. The trick is to find that place where you know you pushed yourself to the limit, you gave your all, as athletes like to say, and be both proud of what you have achieved and eager to push yourself a little harder next time. That's how we find balance between self-pride and humility. Self-pride for valuing the hard work that got us where we are, and humility in understanding that there is always room for improvement. Too much self-pride and too little humility leads to arrogance; too little self-pride and too much humility leads to low self-esteem and low achievement.

## The Immigrant and the Two Frogs

It was midafternoon after the casting lesson, and we still had a couple of hours in the mountains before returning to the city. I asked Alexandre if he minded driving me around a bit, so I could check the area in more detail. I opted against riding a horse, after a previous experience had left me unable to sit for a couple of days.

We took off toward the main attraction of the Silveira valley, the Cachoeirão dos Rodrigues (which translates as something like the Big Rodrigues Waterfall), a wide drop of some fifty feet in the river, allowing for amazingly diverse types of waterfalls, from thin rivulets to massive downpours. Around the waterfall, Nature seems to explode in exuberance, the high tropical forest and scattered pastures combining with pine trees to produce a mosaic of countless shades of green.

There was no one in sight. Only a few cows grazing here and there, and the incessant roaring of the waters. I thought of the stark scenery surrounding the Cow Green Reservoir in Cumbria, the austerity of the New England winters, and felt a nostalgic pang for the country where I was born and lived for the first twenty-three years of my life. I had been away for longer than that now. Where was home? Brazil? The United States? What *is* home, anyway? An immigrant is a perpetual searcher: for the old home of his childhood and a new home for the present. The roots, the extended family, the childhood friends, the familiar places from childhood, exist only in your head as floating, fading memories.

With each trip to your home country you feel more like a tourist, like someone who slept for years and woke up old and disconnected from his past, like Rip Van Winkle. The old friends are unrecognizable, the familiar places either don't exist or are so changed as to have become unfamiliar; time and distance have uprooted your past. You walk around the streets of your childhood looking for clues, knowing that the people you used to greet everyday, the doorman, the ice cream vendor, the corner barber, are all dead; that the trees you used to climb have been cut down to accommodate a larger traffic flow; that the shops are all different; that you are different. You feel dispossessed of a part of yourself, robbed of your past. There is a gap in time, a book with the middle chapters missing. There is no cemetery or museum for your past. I feel so down around my old neighborhood in Rio that I have chosen not to return there. Better to hold on to the little of my past that my memory can preserve. At least what I remember is still alive in me, even if distorted.

If Rio is not my home anymore, where is my home? The answer is surprisingly simple: it's where I am now, with wife and children. Hanover, in New Hampshire — a state I hardly heard of growing up — has been home now for twenty-four years, longer than I lived in Rio, even if the memories from here are less intense. (Even New England was a very vague reference for a boy growing up in Rio de Janeiro, although my parents and two older brothers spent two years in Boston in the early 1950s while my father pursued a master's degree in dentistry at Harvard.) I planted new roots, coming alone to the United States at twenty-seven. There was no community of immigrants to be part of, no religious groups, no family; only my fellow physicists and co-workers at Fermilab, the huge particle accelerator forty miles west of Chicago. I made this country my own, where I built my professional life and created a new lineage of the Gleiser family with my five children. I am deeply attached to my life here and wouldn't change it for anything. Still, every time I go to Brazil and hear the *bem-te-vi* singing in the palm trees and the Southern Cross shining bright in the night sky, I feel the same nostalgic pang I felt at the Silveira valley that afternoon. From the perspective of a human life, the bird and the stars are timeless — they were there before me and will be there after. I now realize that my pain is not from missing the past; it's from missing the future, from knowing that time is marching forward and less is to come. I look at my wife and children and realize that every new day I have to love them is one fewer day I have to love them.

Faced with loss, as everyone is, immigrant or not, we have two choices. Wallow in self-pity and spoil the short time we have; or celebrate life, making each day matter. Even if some days I struggle to light up the candle, I have vowed to embrace the latter option and, as the Welsh poet Dylan Thomas expressed so well, to "rage, rage against the dying of the light." Thomas wrote this poem to his dying father, to give him strength to fight his illness. "Do not go gentle into that good night."

I wish I had said this to my father, who pursued his night with a passion, smoking cigarette after cigarette for years, each puff whittling off a bit more of his soul. He was in a hurry to meet my mother. "It's time for me to join her," he told me, a few months before dying of pancreatic cancer. Love can do that, it can distort our sense of purpose, making the promise of the eternal that much more real and comforting.

It's sadly ironic that this is the same man who, when I was a troubled little boy uncertain of the future, would sit me on his lap and tell me the tale of the two little frogs and the pail of milk, a tale I now tell my children.

"Once upon a time, two little frogs fell into a pail filled with milk. The first little frog paddled and paddled, trying to jump out of the pail. But the pail was high and his legs small. Soon he got tired and gave up, drowning right away. The second little frog was different. He paddled and paddled, just like the first little frog. But he wouldn't give up, even as he got tired. He kept on paddling and paddling, paddling and paddling. So persistent was he that he churned the milk into butter and, with great effort, was able to jump out of the pail."

After he was finished, my father would look at me with his dark brown eyes, knowing well the effect the story had on me. "So, Marcelo, what kind of frog do *you* want to be?" I've been paddling ever since.

# 3

# Sansepolcro, Tuscany, Italy

## Michelangelo's Trout

By 2008, I was deeply into both fly-fishing and a whole new research direction, the origin of life. My passion for fly-fishing had
grown to such an extent that winters in New Hampshire became
painfully long. Like local motorcycle riders, who store their bikes
for five to six months a year (I look longingly at our Triumph
freezing in the barn from November to April), we can fish only
between mid-April and late October around here. Some diehards

do venture out to the rivers during winter, and inspired by these brave souls I tried it a few times. Alas, my attempts were, to put it mildly, not rewarding. Between getting my line cut by shards of floating and partially submerged ice, and fingers so frozen that tying a fly took almost five minutes, I decided I might as well just wait. "Delayed gratification," I kept saying to myself. "When the time comes it will be even better."

I love what people say up here: "If you don't do winter, winter will do you." But doing winter is about practicing winter sports, meaning, to most people, skiing, skating, and snowshoeing. With apologies to enthusiasts, ice fishing is not a sport. And fly-fishing, at least to me, is not a winter sport. So I waited, and appeased my urge by reading books about fly-fishing, from the old classics like Walton's to Rich Tosches's hilarious *Zipping My Fly* and David James Duncan's inspiring *The River Why*. I considered taking up fly tying — making my own flies — but quickly realized I didn't have the time for it. Sometimes life does get in the way.

As I was eagerly counting the days until fishing season started (usually after the rivers calm down from the winter melt, around or after April 15), I got an invitation to go to Florence, for a conference organized by the International Astrobiology Society (ISSOL). The acronym is historical. When founded in 1973, the society was called the International Society for the Study of the Origin of Life, as its members were mostly concerned with understanding the origin of life on Earth. However, with the spectacular advances in observational astronomy, the converging interests of several scientific disciplines — from earth and atmospheric sciences to biochemistry and genetics — and the financial backing of NASA and the European Space Agency, investigating the existence of alien life became a sanctioned and fundable research topic: it is possible to get a grant to study the origin of life on

Earth and the possibility of life elsewhere from a variety of points of view.

The combination of going to Tuscany, one of my favorite places in the world, and a conference with the world's top experts in astrobiology was unbeatable. But was there any fly-fishing around Florence? Again, the answer was easy to find on the Web. "Fly-fishing in Tuscany" took me straight to Luca Castellani, a guide in Sansepolcro, a small town in the province of Arezzo, where Tuscany bounds Umbria. What a treat to experience top-notch science and fishing at the heart of the Renaissance.

"You go to the pensione Podere Violino, also where our club is," wrote Luca. "The Mosca Club Altotevere. Best way is to take a taxi from Arezzo."

The plan was to spend two days fishing at what the Italians call the Tail Water Tevere, near the birthplace of the famed Tiber River (Tevere in Italian), the same chalky waters that cut right through Rome. Considered one of the top ten fly-fishing spots in Europe, the tailwater runs from gates at the bottom of the Montedoglio Reservoir, an arrangement that keeps its waters permanently cold.

Fly-fishing in Tuscany was an altogether new and welcome experience for me, fusing the existential and the lyrical in unique ways. To the members of the Mosca Club (*mosca* means fly in Italian), fly-fishing is a bridge between the region's present and the venerable past, where much of the modern world started to take shape. The birthplace of the master painter Piero della Francesca and other luminaries from the Italian Renaissance, Sansepolcro is not too far from Caprese, where Michelangelo was born in 1475. Piero's masterly fresco *Resurrection*, finished around 1460, which Aldous Huxley in a 1925 essay called the world's "best picture," still adorns the walls of Sansepolcro's Museo Civico. Christ emerges victorious from death, his body athletic, left foot resting atop his sarcophagus, right hand holding a flag in military fashion, eyes staring vaguely at a distance, as if facing the divine realm. The painting is unusual in that it contains two vanishing points, the point (or points) in a painting or drawing where lines seem to converge so as to create a sense of perspective. In *Resurrection*, the two points are strategically positioned to separate the human realm (four guards are asleep in the lower part of the painting) from the divine, symbolizing the two dimensions of being.

Tuscans are justly proud of their heritage. What delighted me was to see how the members of the Mosca Club combined history and philosophy with fly-fishing to create an experience like no other. As an illustration, here is an excerpt from the club's website (my translation):

> To fly-fish is to come together, to seek unity. You feel the pebbles vibrating under your boots, the nymphs scrambling, hitting your waders. You breathe with the gills of a fish, and you feel the fear, the recklessness of the moment. Fortunate is the fisherman who can free his senses from the grip and formalities of life, for he is one of the elect.
>
> In the water, a fisherman becomes one with his craft only when he blends archetype and memory. A great fisherman is at once romantic and medieval, a Flemish painter, an Italian from 1200, a lover of the imaginative dimension that defined the Renaissance; he is heart and intellect.
>
> He can see, feel the signs, the iconography of those who long before him made passage through these waters.
>
> Just watch and see. Hear them . . .

Where else but at the beating heart of the Renaissance, that explosive celebration of human creativity, can this alliance between fly-fishing, culture, and our search for meaning take hold? To Luca and the fellow members of the Mosca Club, fly-fishing is an act of reverence, of devotion. Devotion to history, to Nature, to culture, to freedom. I was captivated immediately, knowing I would be among a group of people who took fly-fishing to a new level of sophistication.

Luca is a tall, elegant man, who lives and breathes fly-fishing. He welcomed me as Italians do, effusively, sincerely happy to see me, knowing that he would soon be taking me to his temple in the waters.

"Benvenuto, Marcelo! But you are not Italian, right?"

"No, my Marcelo is the Brazilian version, with one *l*. A very popular name there; there were three Marcelos in my elementary school classroom of twenty students."

"But your Italian is great, bravo!"

"I love it. It sings, just like Brazilian Portuguese."

"You will be singing tomorrow, when you step into our beautiful Tevere!"

"Can't wait."

"But first, let me take you to the club."

The headquarters of the Mosca Club Altotevere are located on the bottom floor of the Podere Violino building, facing the beautiful grounds. A room filled with fly-fishing memorabilia, posters, walls covered with pictures of smiling members holding their trout, of great lunches with the whole gang. This was a tribe I wanted to belong to.

"Look over there," said Luca, pointing to the wall next to a large window. "These are the best flies from last year's fly-tying championship. What do you think?"

Glassed inside a wooden frame was the most exquisite collection of flies I had ever seen. A profusion of colorful feathers, bright blues, greens, and reds, adorning hooks, each was a work of art. One looked like two iridescent-blue butterflies attached to each other by the head; another, like a peacock wearing a golden crown. It was fly-fishing meets the exotic masks and plumes of the Venetian Carnival. Mostly, these were not functional flies; they were objects of pure beauty, to behold, to venerate, bringing together the best of centuries-old Italian craftsmanship, celebrating the elegant symmetry of design, as offerings to the river, to the trout.

"The brown trout here, we call it *trota michelangelo*, since the great painter was born in a town near Sansepolcro. When you see it, you will agree that it is the most beautiful of all trout, as if the master himself had a say in its design."

I could hardly sleep. I had a few too many glasses of Rosso di Montalcino and stuffed myself on *ribolitta*, the absurdly delicious Tuscan soup made with two-day-old bread, kale, cannellini beans, and carrots, all doused with extra-virgin olive oil and served with chopped raw onions and grated Parmigiano-Reggiano on top. Just writing about it makes me want to go back there for more. My stomach, like my head, was happy but bloated, filled with the anticipation of stepping into the same waters where Petrarch, Michelangelo, and Piero della Francesca washed their feet. I could see them strolling along its green banks, soaking in the inspiration from its golden waters.

The next morning, after the obligatory cappuccino and biscotti, Luca took me to the water with another member of the club. He gave me a 5-weight rod fitted with a very strange-looking fly at the end of the leader, a far cry from the gorgeously plumed-up winners of last year's fly-tying championship. It looked like a gray rubber peg with some yellow adornment stuck in the middle. "I know, it's very *brutta, la poverina*. But here this funny thing works very well," said Luca, reading my surprise. "I wonder if it would work in your river? You must try and let me know."

"Of course," I replied, marveling at the surroundings. The Tevere here is narrow, no more than twenty-five yards across, with golden green waters that flow at many different speeds and depths, perfect for trout to hide, waiting for their food to drift from upstream. Hard to believe that this is the same river that runs right outside the Vatican walls.

"Cast upriver and let it drift," instructed Luca. "Make sure you don't shoot the line too far, or it will get caught in the brambles over the edge."

Of course, that's exactly what happened for the first two or three casts. Luca smiled, used to clumsy first-timers in these waters. "After you catch a few here, I have a surprise for you."

"Okay, I'll get this now." With extra focus, I managed a few good casts, making sure my casting arm was close to my body. I took a few steps into deeper water, eyeing a big fallen tree by the shadowy side of the riverbank. You don't ever want to project your shadow in the direction of your cast, as trout can see this very clearly and will scram far away, taking you for a bird of prey. Yes! The rod bent down in an elegant arch, and I greeted my first *trota michelangelo* on the other side, no doubt upset for having been fooled by a foreigner with a fake Italian name.

Luca was the gentlest of guides, always there but never too close. He gave me space, knowing that a true fly-fisherman loves solitude above all. At least he intuited that this one did. I like being on my own in the river, away from everyone, preferably unable to see anyone around. You and the water, the sound of it rushing against your boots, of insects buzzing around, everything bathed in a magical golden radiance, the same that adorns the head of countless paintings of the Madonna holding baby Jesus in churches and monasteries around these parts.

While I labored to catch a handful of trout, Luca's friend must have caught fifteen with disconcerting ease. Every one of his casts brought a trout back with it. I was stunned.

I looked at Luca in disbelief. "How does he do that?" I asked, shaking my head.

"Oh, he comes here a lot," replied Luca. "Knows all the best spots. He has a *testa di trota*, you know," meaning the guy could think like a trout. "Don't pay attention to him. Focus on your own experience."

Something about this high-efficiency fishing bothered me. Doesn't it become pointless? If one is always victorious, what's the fun of competing?

Following Luca's advice, I doubled my focus and moved downstream to a more secluded spot. A short cast along a large boulder produced an absolutely spectacular creature, shining gold with bright red dots along its body, darker tail fin for contrast. This one snatched a bright pink fly, as if in search of something to wear to the underwater ball that night. Its imperfect symmetry was an expression of pure divine beauty.

"Okay, now you can have my surprise: I will bring you out tonight, so you can experience fly-fishing in the dark," Luca told me, admiring the trout. "Here, let me help you so we don't hurt her." I have never encountered, before or since, a fly-fisherman so careful with the trout. Sure, most practice catch and release, as they should. But Luca went overboard, making sure the trout was okay, that the damage was minimal. It was almost as if he could feel it in his lips. The hooks were barbless, which made it extra challenging to keep the fish hooked during battle. But it also made for a fairer and more exciting experience. You lose more fish, but know that the ones you net are due to your skill and not to a nasty metal trap.

After a huge dish of *pasta e fagioli* and less wine than the previous night we were out again, under a moonless sky. Headlamp on and rod in hand, Luca took me to a shallow and wider part of the river. "Try to cast across the water and let it drift; at night you don't have to be so careful with how the line hits the water or if it's causing a little tail. The trout half sees the action and goes for it." *

I was initially uncomfortable with the whole thing, afraid I was going to make a fool of myself getting horribly tangled up. At the same time, I was very curious to see if it worked. To my complete surprise, after a couple of hours I had hooked four trout and brought in two, not bad for a rookie night fisherman.

"The beautiful thing about fishing at night is that without your eyesight to guide you, you must fish by instinct," said Luca. "It forces you to be one with the water, with the rod, with the trout; no distractions. It's pure poetry."

As we were walking back, I felt myself to be on a different plane of being; something had snapped inside, as I pushed myself beyond what I thought I was capable of. I finally felt that I could hold my own fishing pretty much anywhere. I had much to learn still, but I wasn't scared anymore.

The boy looked at me from across the river, waving for me to go on. I took a few more steps into the monastery. I could now see the altar, candles and incense burning, a large golden pillow on the floor inviting me to sit and contemplate. I stared at the candles and incense for a while, mesmerized. The flames flickered bright,

---

*A tail means a little turbulence as the line streams along, fighting the direction of the current, something you must avoid during daytime, lest you spook the fish.

but didn't consume them. Remembering Moses and the burning bush, I waited for a voice. But there was only silence. If God was there, she chose not to speak. Maybe the river was her voice.

"Tomorrow, I will take you to my secret spot," said Luca. "You deserve it."

That night, the boy came to me in my dreams. His head glowed with the same golden light of the Tuscan sun. This time, though, he wasn't alone. My mother held his hand. She wore a long blue headdress decorated with golden stars, like an early Renaissance Madonna. They floated down from the hills toward me, smiling. To their left, the trees were bare; to their right, they were in full bloom. The sun too was half bright, half dark. My mother slowly approached me and kissed me tenderly on the forehead. I was six the last time she kissed me. She had watched me today; she had always watched me. I looked up, drinking in the brightness of her eyes, keeping it inside me, alive in me. Still smiling, my mother and the boy turned around and, without saying goodbye, floated back to the distant hills they had come from.

When the alarm went off in the morning, my eyes were glued shut. I had cried in my sleep.

## Baby Earth, Baby Life

After my immersion in blissful fly-fishing, it was time to put on my scientist hat and join the conference in Florence. With more than 350 scientists from all over the world, the meeting would put me in touch with the latest ideas about the origin of life on Earth and the possibility of life elsewhere.

Consider life as we know it. You hike through a forest, or dive in a coral reef, or watch a David Attenborough BBC documentary, and the variety of living forms is mind-blowing. Now put this in the context of our planet's history, a small sphere formed some 4.54 billion years ago from leftover matter that didn't make it into the Sun. How did Earth go from a nascent lifeless planet to this crucible of life? Why is there life here and not, say, on Venus or Jupiter? How did dead atoms become thinking molecular machines?

In the beginning, what was to become the solar system was a giant floating cloud of mostly hydrogen — the simplest and most abundant chemical element in the Universe, with a single proton in its nucleus. This hydrogen cloud was sprinkled with a bunch of heavier elements, including carbon, oxygen, calcium, gold, iron, etc. Where did all this stuff come from? We now know that only hydrogen and helium, the two lightest chemical elements, emerged during the cosmic infancy, well before stars existed. Other primordial light elements were also synthesized early on, including lithium and deuterium (a kind of hydrogen with one proton and one neutron in its atomic nucleus, what we call an isotope), but in much smaller quantities. To a cosmologist, the chemical elements that matter in the Universe are hydrogen (75 percent of the total) and helium (24 percent of the total). Everything else — the other elements of the periodic table — was synthesized much later, by dying stars. When scientists say that we are star stuff, they actually mean it. We are moving, living bundles of stardust, driven to ponder our origins, animated remains of dead stars, cosmic nuggets of thinking matter. As we look up at the heavens we see our own past; perhaps, in some way, we long to establish a connection with our ancient cosmic roots. Whoever thinks that science and poetry are incompatible should ponder this.

Like us, stars have life cycles, being born, evolving, and finally dying, after they expend all their fuel. To a star, however, the fuel doesn't come from the outside. The fuel is its own innards: to exist, a star needs to consume itself, to self-cannibalize. Being essentially giant balls of hydrogen, stars produce enough radiation pressure to sustain themselves against their own demise: a final implosion by gravity. This comes at a cost, as stars must burn their own hydrogen — convert it into helium in a process known as nuclear fusion — to fight the inexorable inward pull toward their doom. After a while — usually billions of years — they run out of hydrogen at their core and must start to burn helium. Helium is then converted into carbon and some oxygen. Depending on how massive the star is, the process may go on to fuse heavier elements, or stop there. At a certain point there is a huge bang (or more than one, details being more complicated than this), and the star spits most of its entrails into outer space, sprinkling it with the heavy chemicals needed for life.

Imagine then a slowly spinning hydrogen cloud floating around the galaxy some five billion years ago. At some point, cataclysmic shock waves from surrounding dying stars shook it out of its peace, seeding it with heavy chemicals, while gravity caused it to shrink. As the cloud shrank and spun increasingly faster, it gradually assumed the form of a flattened disk as most matter fell to the center, while leftover stuff circled chaotically around it. With time, the central stuff got dense and hot, finally igniting into our Sun, while the rest coalesced into different worlds, of varying size and composition. In less than a billion years the solar system was born, with Earth the third planet from the Sun. What happened here happens over and over across the vastness of space. Dying stars spawn new ones, a cycle of creation and destruction that replicates itself throughout the Universe. Nature is in a constant state of flux; energy flows and matter dances from pattern to pattern.

Cut now to the nascent Earth, a hot ball of molten matter, bubbling and boiling as it slowly took shape. Debris from left-over matter in the form of comets and meteors collided merci-lessly with young Earth, bringing their own treasures: water and a variety of chemicals, including some simple organics, chemical compounds with carbon. We could call this the angry gifting era. The onslaught would quiet down only some 600 million years after Earth's origin, around 3.9 billion years ago. If elements had gathered to make some kind of living creature before then, it was probably destroyed without a trace. Life may have had many false starts, lost in the dust of time. To understand the history of life on Earth we must understand Earth's life history. This is true for any planet that may harbor life, past, present, or future: *the history of life on a planet is contingent on the planet's life history*. And given that no two planets have the same history, life is not a repeat-able experiment. Life may share the same biochemical principles across the Universe, being carbon-based and gene-driven by Dar-winian natural selection. But each world will harbor its own kind of living forms, which will mutate and evolve according to chance and that world's unique changing environment.

The preceding argument speaks to us directly. Life on Earth is unique: if there is life elsewhere, it will be different. And given that we are the most sophisticated life form on this planet — I do love and respect whales, dolphins, monkeys, dogs, cats, etc., but I'm talking about reasoning ability — we are unique as intelligent beings. If there is any kind of intelligent life form in another stel-lar system (more on this later) it will not be like us, even if it may share some of our traits, such as an approximate left-right symme-try. Furthermore, any alien life form will be so far away that, for all practical purposes, we are alone. This means that we matter in a very central way, what I like to call *humancentrism*.

The next time someone says that we are nothing in the vast-ness of the cosmos, that science teaches us how worthless we are, please beg to differ. We matter for our uniqueness as thinking mo-lecular machines, in a planet that matters for its uniqueness as a stable harbor for the prolonged existence of life. These two inter-related conditions—the existence of intelligent life and a planet that offers long-lasting shelter for its creatures—are not likely to be encountered with high probability in the cosmos.

Back to life's origin here: what we do know with certainty is that it had taken hold at least 3.5 billion years ago, in the form of very simple one-celled creatures known as prokaryotes, organ-isms with neither a distinct nucleus housing genetic material nor specialized organelles, such as the mitochondria found in more advanced cells. These primitive living entities were identified in stromatolites, rocklike formations found in Western Australia that look like layer cakes. It is possible that life took hold even earlier than this, although recent claims have not been substan-tiated. Even so, since the late heavy bombardment ended around 3.9 billion years ago, if life appeared "only" 400 million years after, we may consider it to be a fairly fast development, given its complexity. This bodes well for finding life elsewhere, in worlds with the right conditions. But lest giddy optimism make us over-confident, it's good to keep in mind that there is an enormous jump from simple carbon compounds, even simple amino acids, to a living one-celled organism. Life needs millions of complex carbon-based molecules to act in tandem, generating the meta-bolic output to keep it going while protecting itself from a harsh outside environment and predators. On top of that, life is viable only if it reproduces. Otherwise, life would last only as long as a single creature. In fact, without reproduction, the concept of a species is meaningless. At their bare minimum, living creatures are molecular machines capable of reproducing according to the principles of Darwinian evolution and genetic natural selection.

Of the three key questions about the origin of life — When? Where? How? — the "when" question is in a sense the simplest. Although we can never be absolutely sure if the earliest sample of life we have discovered is the earliest (it's always possible to find an earlier one, at least until we hit the age of heavy bombardment), the question relies on searching and identification methods. Even if the biochemistry and geochemistry to identify traces of our distant ancestors in early mineral samples are exceedingly complex, there is no fundamental conceptual block impeding progress. On the other hand, the "where" and certainly the "how" questions are a different matter, even if they are all related.

There is no consensus as to where the first living forms appeared. Darwin wrote of "warm little ponds," hinting that somehow shallow water was the ticket. His intuition may well have been correct, as complex molecules need a liquid medium to drift about and find one another, in order to react and connect. Also, well before there were continents, early Earth was mostly covered by a vast shallow ocean. Ponds susceptible to tidal variations in depth and temperature may have provided ideal conditions for prelife chemistry to cook up living chemistry, a possibility I explored with my then graduate student Sara Walker, now a professor at Arizona State University. However, alternatives such as clay materials (basically, mud) or underwater hot thermal vents are also reasonable, each offering advantages and disadvantages. Life may have emerged at different spots and at different times, instead of having radiated from a single origin. For example, if early life needed a boost from falling debris, such as meteors or comets carrying a necessary ingredient, it may have sprouted at different spots where the right conditions and chemicals found each other.

Given the amazing versatility of terrestrial life forms, persevering and even thriving in the most extreme environments, we should keep an open mind as to where life could have appeared. There may be a variety of answers to this question.

The answer to the "how" question, though, remains shrouded in mystery. The simplest life forms we know, prokaryotes, are already vastly more complex that what probably transpired here 3.5 billion years ago. Unlike with physics, where we can trace down the building bricks of matter by a systematic deconstruction of large to small — from visible stuff to molecules to atoms to elementary particles — with life we hit a dead end, the primitive cell. We can picture a proton and an electron coming together to make up a hydrogen atom, and know how to compute this in detail and even compute when it happened in the cosmic history (about 380,000 years after the "bang"). But a simple cell is already a hugely complex entity, filled with all sorts of different molecules performing distinct functions, surrounded by a protective membrane that, like a nightclub bouncer, allows only some things to go in and out.

Faced with such a daunting challenge, scientists take two complementary approaches: from top down, they consider a simple cell and systematically strip it of genetic material and other molecules to arrive at a sort of essential living thing, the bare minimum of life. With this life unit in hand, the question then becomes how such a complex living thing emerged in the first place. As a second approach, scientists attempt to build life from the bottom up, taking different molecules as Lego blocks and assembling them with increased complexity until the transition point is reached and the assembly becomes a living entity. The Florence meeting had prominent scientists describing their progress on both fronts. Gerald Joyce, from the Scripps Research Institute, presented his mind-blowing experiments with self-assembling compet-ing chunks of RNA that illustrate natural selection on the fly.

The essential difficulty with either approach is that even if an experiment is successful, we can never be sure that this was the path that life took on primal Earth. Amazing as it would be, the ability to create life in the laboratory would not answer how life emerged here over 3.5 billion years ago. The best that we can aim for is to construct viable scenarios, hoping that they illuminate, to some unknown degree, what happened here and may have hap-pened elsewhere in the cosmos. We can't go back to baby Earth and extract all the information needed to understand how life first emerged. What information we can gather through our cre-ative methodology and diligence is necessarily incomplete. The *exact* details of Earth's past, of the unfolding biochemistry in its primal pools and soil, are unknowable to us. Even if we were so lucky as to spot an Earthlike young exoplanet where life is about to emerge, the details of what transpires there may approximate but would never match what has transpired here.

We are thus led to an important conclusion: we will never know for sure how life originated on Earth. Unless a rigorous proof is offered that life can have only one or a small number of biochemical pathways, the specifics of the origin of life anywhere are unknowable, a mystery that will remain with us for the fore-seeable future, if not forever.

Some people react negatively to such statements of what sci-ence can and cannot achieve. They shouldn't. It's important to bring the workings of science — both its endless potential and its inherent limitations — out in the open. There is much inflated rhetoric about what science can accomplish, reflecting a kind of old-fashioned positivist triumphalism that science can conquer all. A more realistic view is to consider science as a human enterprise and, as such, as limited and fallible as we are. To convince oneself of this, all it takes is a quick look at the history of science, where concepts and worldviews continually evolved as we learned more about the natural world. An honest view of science doesn't take away from its beauty or tremendous power. The fact that we can't get to a final answer doesn't mean that we should give up trying to understand all that we can about the question. As I have argued before, science is not about final answers; science is about constructing better descriptions of Nature based on more efficient data and modeling. To this gradual buildup we must add limitations to what knowledge we can acquire, limitations that come from Nature itself: the finite speed of light, which precludes us from gaining information beyond our cosmic bubble; quantum uncertainty, which adds a fundamentally unpredictable element of randomness at the very core of matter; our incomplete view of the human brain and how it engenders mind. Taken together, these limitations hinder our grasp of the ultimate nature of reality. At the same time, it's important to stress that limits to knowledge are key to the progress of science. To overcome an obstacle we must first face it. The question of *knowability*, then, becomes central to our understanding of the world and of our place in it.

## Knowledge, the Endless Pursuit

When we look at the world around us, we don't see everything. We can't. Humans evolved in a very specific environment and adapted to maximize their chances of surviving in it. We are creatures of the Earth, a planet bathed in radiation from a star with a surface temperature of about 6,000 degrees Celsius (about 10,800 degrees Fahrenheit). A star's surface temperature determines the kind of radiation it emits the most, what we call its peak power. What gets down to Earth's surface, after the atmosphere filters and scatters it, allows animals to ultimately fulfill their two most essential needs: to eat and to reproduce. It is thus no wonder that most surface animals see in what we call the "visible window" of the electromagnetic spectrum, roughly the colors of the rainbow. Those frequencies correspond to the Sun's peak output. Some species see in infrared and ultraviolet wavelengths as opposed to the visible portion of the electromagnetic spectrum, or use smells to guide them; others, like bats, use echolocation to move about. But most ground-dwelling diurnal species use visible light. Successful life forms are those that can best make use of the resources around them so as to optimize the survival chances of their reproductive output. Humans are no exception.

Given our evolutionary history and the planet we live on, we capture that fraction of reality that is most useful for our survival. This means that there is a whole lot around us that we don't see or know exists. But the fact that we can't see or perceive this part of reality doesn't make it less real. Quite the opposite; as every fisherman knows, these invisible realms are where the excitement is. That's where the possibilities of growth are, where what seems impossible in our immediate reality may actually not be. Humans have an urge to explore the unknown, what lies beyond their immediate reach. This may be our species' most distinguishable trait. Animals want to be safe, living within familiar boundaries that don't expose them to any extra risk. They keep to their tried, well-adapted behavioral patterns, a recipe that allows them to thrive. Even migrant animals are not explorers: deviating from their roots can be deadly. Humans, on the other hand, have a need to lunge into the unknown, to expose themselves to what is uncomfortable, even threatening. We take risks as individuals and as a species, continually pushing ourselves beyond established limits. We like our boundaries elastic, safe but expandable.

As with the physical, so with the mental. We take risks with knowledge, with ways of expressing our thoughts and feelings. Science is much more than a mere description of the natural world; it's our commitment to exploring the unknown, an expression of the very human urge to constantly redefine who we are by expanding our realm of existence. As our view of the world changes, so does our understanding of our place in it and the meaning of our humanity. In this, science joins the arts as an expression of what is most precious to us, our search for meaning.

Circling back to the limits of science and knowledge in general, we can see why they are essential. They are the gateways to invisible realms around us. If we can't see tiny microbes or faraway stars with the naked eye, we invent microscopes and telescopes to expand our vision. If we can't see inside our bodies or underwater, we invent X-ray machines and sonar to expand our vision. What we call reality is constantly shifting, as we expand our reach into the unknown. As long as scientists have funds to keep exploring, there's no end to this pursuit. And that's how it should be. We need the thrill of pursuing new knowledge. We want to push those elastic boundaries of our perceived reality outward. The real defeat is to believe that there is an end to this quest. Imagine how sad it would be if we one day arrived at the end of knowledge. No new fundamental questions to explore, no boundaries to expand, no great discoveries ahead. Only tiny adjustments here and there to what is already known. Remarkably, many scientists and thinkers have speculated that this would indeed happen, that one day we would arrive at the end of knowledge. Some even declared we were there already. I am thus glad to report that these people were wrong. Given the very nature of knowledge, every discovery starts in ignorance. New discoveries may answer a few questions but invariably create new ones. Indeed, the more fundamental the new discovery, the more fruitful it will be, as it will open doors that we couldn't even have contemplated beforehand. Knowledge starts in ignorance and generates new knowledge that generates more ignorance. Such is the wondrous nature of knowledge, the endless pursuit.

I can't resist comparing this to trout fishing, another endless pursuit. We can't see what's underneath the surface. We go to the water knowing that each new experience will be different, that even for an expert there is always room for surprises. There are new trout to catch and those that are uncatchable; different rivers to explore; different conditions to try out. Every river tells a different story, and that story changes every day, as Heraclitus so wisely understood. Like Jorge Luis Borges's "Library of Babel," which collects all possible books that can be written but is necessarily incomplete, given that in an "indefinite and perhaps infinite" library a complete catalog cannot contain itself or even be defined, there is no final catch, no final rule that explains it all; there is only the daily progress, the daily adventure, small fragments of an endless puzzle that we put together here and there. Nature fills us with wonder and humility. As we try to make sense of it through science, or engage with it in a river with rod in hand or running up a mountain trail, we learn that what we grasp is but a tenuous thread connected to an invisible wholeness. Borrowing again from John Muir, "When we try to pick out anything by itself, we find it hitched to everything else in the Universe." The known, the unknown, and the unknowable form an indissoluble wholeness to which we are hitched.

## Is Anybody Out There?

No one who ponders the origin of life can leave aside the question of alien intelligence. Whenever I give public talks, the question always pops up: Are there intelligent aliens in other worlds? Or are we alone? The question is of huge importance, of course. Some say that the confirmation that alien life exists, especially intelligent alien life, would be the greatest discovery of all time. Although I'm not sure I agree with the hyperbole, the discovery of any kind of alien life — intelligent or not — will surely have an enormous impact in our culture. It will redefine the way we think about ourselves on many levels, affecting everyone, from atheist to believer.

Let me first make an important clarification. When thinking about life in other worlds, we must make a clear distinction between living creatures and *intelligent* living creatures. Most people imagine that if a planet (or moon) has life, either it will have intelligent life or it will develop it in due course. This position assumes that life leads to intelligence — that is, that Darwinian evolution implies that intelligence is an unavoidable consequence of life, that once the seed germinates, it will evolve into a few intermediate critters and, in the fullness of time, bloom into an intelligent creature. That this happened here explains why so many people think this way. After all, intelligence does give its bearer several evolutionary advantages. For a worrying example, we, as the dominant species in this planet, could easily kill all the extant tigers, collecting their hides as trophies. Using our intelligence, we could eliminate any potential animal threat to our survival. (No one said that intelligence and wisdom are the same thing.) Given that what life mostly wants is to reproduce, wouldn't intelligence always be the end goal of the evolutionary game?

It would not. Life is an experiment in adaptation by natural selection. It has no end goal, no final plan. (In other words, life has no teleological purpose.) If living creatures are well adapted, mutations will be mostly lethal or useless. Case in point, consider the only example we have, life on Earth. For about 3 billion of the 3.5 billion years that life has existed on our planet, it was mostly very simple, composed of unicellular creatures. Even so, there was a huge transition in complexity already within one-celled animals, from prokaryotes to eukaryotes. Prokaryotes, as we mentioned before, are living cells where the genetic material is free; eukaryotes, on the other hand, have a protective sack containing their DNA, as well as several small organelles, specialized functional structures. After accounting for this essential and not fully understood jump in the complexity of one-celled organisms, life on Earth was still comparatively simple for most of its history.

The turmoil began when the atmosphere became gradually oxygenated, thanks to the photosynthetic action of our one-celled prokaryotic ancestors. It is amazing that we — as well as all other multiorgan animals — owe our existence to the accidental mutation(s) that led one-celled bacteria to consume the abundant carbon dioxide of Earth's early atmosphere and expel oxygen. Through their action these creatures transformed our planet, giving it a much-needed ingredient to sustain complex life. Current measurements show a rapid growth in atmospheric oxygen concentration starting at about one billion years ago and peaking around five hundred million years ago, a time when multicellular life exploded in diversity, during the so-called Cambrian Explosion. Without the added metabolic advantage of abundant oxygen, chances for complex life would probably have been slim.

We have just encountered one major stumbling block for the emergence of complex life as we know it: oxygen abundance.

Could it have happened elsewhere? Surely; other planets could have oxygen-rich atmospheres. Can we be *sure* that it happened elsewhere, even if the world had abundant prokaryotic forms of life? Absolutely not. Recall that life evolves through accidental mutations. By *accidental* we do mean random, purpose-free changes in the genetic code of a species that could then be transmitted to future generations. In other words, we shouldn't expect that prokaryotic bacteria in other worlds would undergo the same sort of mutations that their relatives experienced here. We can't even be sure that their genetics would work the same way.

Many scientists (especially astronomers and physicists), when faced with the small odds for life and complex life, like to cite the large numbers of worlds: "Just think that in our galaxy there are some 200 billion stars, most of them with planets and most of these with moons. So, in our galaxy alone there should be trillions of other worlds, each unique." This is definitely true, as we know from current searches for exoplanets. "Then extrapolate it to the whole known Universe, where there are hundreds of billions of galaxies, some smaller and others bigger than the Milky Way. Shouldn't we expect that in many of these countless worlds life would have emerged?"

Indeed, to expect is all that we can do, at least for now. The wonderful thing about science is that it brings the unreachable within reach. We may never, or not for thousands of years, be able to travel to a distant world five hundred light-years away from us.* But we can still study that world, collecting information about it right here on Earth and in orbiting telescopes, such as the Hubble

---

*For perspective, our galaxy has a diameter of one hundred thousand light years. Andromeda, the nearest galaxy to our Milky Way, is about two million light-years away.

Space Telescope or its successor, the James Webb Telescope, whose launch is scheduled for October 2018. Like the wise fisherman who knows where best to cast for fish, we can use what we know of exoplanets and their parent stars to guide our search for possible harbors of life, planets that some astronomers like to call Earthlike, that share some of the properties of our cosmic home. In the not-too-distant future, we might be able to collect data from the exoplanet's atmosphere and search for chemical elements that signal the possibility of life, such as water vapor, oxygen, and various carbon compounds. Of course, finding any of these is not a confirmation of alien life. But at least it would indicate a favorable environment. For *confirmation* we would need something more dramatic, like the identification of chlorophyll, the green pigment ultimately responsible for photosynthesis, or, stretching it a bit, the sighting of large-scale engineering projects on the planet's surface, like the Great Wall of China or a giant hydroelectric dam here. Or maybe alien engineering would be space-borne, and we would find artificial moons or a giant orbiting space mirror to amplify rays from a faint star.

Before any of this happens, however, what we can say is that even with the large numbers of worlds out there, and even if many of these might be Earthlike in some respects, life is probably not a widespread cosmic phenomenon. Although we should expect it to flourish elsewhere, given that the same laws of physics and chemistry apply across the Universe, life is almost certainly a rare phenomenon, as we can see from our own solar system, with one data point. (Other candidates, such as Jupiter's moon Europa and Saturn's moon Enceladus, hold a very remote chance. Their scientific value lies not so much in affording potential harbors for life as in offering laboratories to explore different environmental conditions that could lead to life.) Even if life were to be found elsewhere — and I do hope it will be — it would probably be simple and unicellular, given the many biological and planetary challenges for complex life to develop. *Intelligent* complex life, capable of transforming materials and creating technologies, would be a distant possibility. We can't discard it, of course, but we can argue that it would be quite a rare occurrence.

There are many reasons why intelligent life is a long shot. In the early 1950s, the great physicist Enrico Fermi was having lunch with some colleagues at the Los Alamos laboratory cafeteria when, after scribbling a few calculations on a napkin, he exclaimed: "Where is everybody?" His friends looked around, assuring him that everyone was accounted for. "Not you, *ragazzi*," he insisted. "I mean aliens. Where are they?" Fermi estimated that a galaxy like our Milky Way, about ten billion years old and one hundred thousand light-years across, should have been colonized by an intelligent alien species. To see why, consider this: imagine that an intelligent race emerged on a faraway planet about ten million years before we did. Ten million years is peanuts compared to ten billion years. But think how much we have accomplished with only four hundred years of science: from riding horseback to sending spacecraft to probe the edges of our solar system. Similarly, a technologically savvy species with an extra ten million years could have created wonders, inventions that to us would look like magic. As Arthur C. Clarke famously wrote, "Any sufficiently advanced technology is indistinguishable from magic." Assuming they shared our wanderlust, even if traveling at one-tenth the speed of light, 18,600 miles per second, they would have had plenty of time to spread across the whole galaxy, including here. (At that speed, they could cross the entire galaxy in one million years.) So, asked Fermi, "Where is everybody?"

There are many answers to Fermi's question, supposing intelligent aliens exist. They could have self-destructed after discovering nuclear energy; they could have evolved to such an extent that colonization of other worlds became superfluous; they could have visited here already and hated it so much that they left without leaving any clues; they could be here or they could be watching us from a distance using some amazing stealth technology, and we wouldn't have a clue; they could have made us; we could be characters in their video games; they may not share our wanderlust, being smart to stay put in their world, lest some predatory species discover them; and so on. Or, even if they do exist, they are so far away that it makes no sense to them to travel across such enormous distances to visit Earth. The aliens may be content exploring a region around their main star of, say, one thousand light-years, still far away from us here. In this case, they may be out there and we wouldn't know. Even if intelligent life exists elsewhere, we may never — or not for a very long time, since "never" is a dangerous word to use in science — find out.

There probably *is* life elsewhere — but we must come to terms with the fact that, in practice, we are it when it comes to an intelligent species. Until they say hello or come to visit, and discounting the very small odds that we will find evidence of alien intelligence in the next few decades, for all practical purposes *we are alone.* To put it differently, if there is intelligent life out there, we are unaware of its existence and will remain so for the foreseeable future. This conclusion is — or should be — a game changer. It should change the way we relate to one another, to our host planet, and to the other creatures living here. It should chart a new ethics for each of us and for humanity as a whole. It should broaden Kant's categorical imperative to respect other rational beings to include *all* beings and the planet. It should redefine our destiny.

To see how, we need a proper setting. We need to go to one of the strangest and most spectacular spots on Earth: Iceland, land of famed trout and salmon, of trolls and active volcanoes. There the fisherman will find his peace, the scientist his mission, and Earth will guide us in our search for meaning.

You could not discover the limits of soul,
even if you traveled by every path in order to do so;
such is the depth of its meaning.

HERACLITUS

◉

# 4

## Laxá River, Mývatnssveit, Iceland

### You Must Leave This Ship!

Any lover of fly-fishing who suffers from uncontrollable wander-lust wouldn't refuse a chance to go to Iceland. I was all smiles when the message from Dartmouth's Alumni Travel came with the invitation to lead a group of alumni on a cruise circumnavi-gating Iceland. "You are free to choose the best topic for your lec-tures, of course," they wrote. Of course. And what country could be a better setting to discuss the geological history of Earth and the recent controversy on global warming? I could start with the origin of our planet and advance all the way to the Anthropocene —the proposed name for the current geological era, at least in a few circles—the era of humankind and its incontrovertible impact on the globe. And I could add a few extra days at the end of the trip to go after the famed Icelandic salmon or brown trout. Alone, in the Icelandic wilderness, just me and a guide. It would be the greatest test yet of my fishing skills, a pilgrimage into one of the most spectacular and mysterious spots on the planet, with active snow-covered volcanoes, sulfur bubbling from underground alongside steam-shooting geysers, hundreds of dramatic waterfalls, even elves dancing on lichen-covered stones. The belief in *hundúfolk* (hidden people) is so pervasive that construction projects often have to deviate from sites and stones where elves are believed to dwell.

This was a trip to bring along my wife, Kari, then thirty weeks pregnant, and our five-year-old son, Lucian. We couldn't be more excited—a nice cruise ship, easy traveling and easy shore excursions, good company and conversation onboard, perfect for a pregnant lady and a child.

Just to make sure, Kari consulted with her doctor a few days before the trip. "Go and enjoy," she said. "Nothing to worry about. Your vitals are in great shape, and so are you." Kari is a *very* serious amateur athlete. She ran regularly until six months' pregnant and swam until the day we boarded. Nowadays she ranks among the top female obstacle-racers in the world. Six leisurely days aboard a cruise ship were supposed to be a piece of cake.

On July 7, a few days before the trip, I was searching for a topic to write about for my National Public Radio blog on science and culture when I found this news from *Daily India*: "Iceland's Hekla volcano may erupt any time: Geologists are monitoring unusual activity on one of Iceland's most feared volcanoes, which is likely to erupt anytime, raising fears of a new ash cloud over Europe."

Seriously? In the Middle Ages Mount Hekla was known as the "Gateway to Hell" for its unmatched ash-and-lava-sputtering capacity. Monks wrote of its fearsome appearance and power, of witches gathering there at Easter, of souls in the shape of birds being sucked down into its entrails during an eruption. To be named *the* Gateway to Hell is no small distinction on a small island about the size of Pennsylvania covered with 130 volcanic mountains, 18 of which had erupted multiple times since humans first settled there around 900 CE. Hekla alone has erupted over twenty times in this period, the last being on February 26, 2000. A major eruption from Hekla would have devastating consequences. A small one would at the very least ruin our cruise, and my fishing plans. In May 2010, the volcano with the hardest name to pronounce in any language, Eyjafjallajökull, had caused major havoc in Europe with its thick, poisonous ashes.

I wrote to the Dartmouth staff alerting them of the impending doom. Should the whole thing be canceled? We asked the local tour operators for any concerns and monitored the news for the next couple of days. Things seemed to have calmed down. I guess volcanic eruptions in Iceland are taken as nonchalantly as earthquakes in California. People know about them and take precautions (collecting gas masks, pickaxes, and water and food in their homes) but choose not to dwell on it.

We took the quick flight from Boston to Reykjavík, excited to board our ship the next day. We arrived midafternoon and went exploring the neighborhood around the hotel. The city flanks a quiet bay with gorgeous views of snow-covered peaks in the distance, including the amazing Snoefellsjökull, a volcano on the western peninsula of great literary distinction, which inspired Jules Verne in his book *Journey to the Center of the Earth*. The first bizarre impression was the lack of trees. There were a few scattered here and there, but looking out into the mountains all we could see were barren slopes. Touristy stores were filled with Viking swords and helmets and cuddly ugly trolls and elves, mixed with puffin dolls. Time passed, the day ended, and ... night didn't come. Mid-July at such high latitude (64°08' N), the sun traces a shallow arc near the horizon, dipping below only for a short while, creating a prolonged and often spectacular twilight. All homes have very heavy curtains or closable wooden boards on the windows.

The next day we joined the traveling group and transferred to the French-operated ship *Le Boreal*. After unpacking we went to dinner, where we joined a few of the Dartmouth alumni contingent. We were in the midst of a pleasant conversation about the many adventures ahead when an older gentleman tried unsuccessfully to stand up, declaring he couldn't feel his legs. Kari and I looked at each other in slight panic, wondering what to do. "Not to worry," said the gentleman, rubbing his thighs, "this happens often. I'll be fine." His wife rolled her eyes. Better not get mixed up in this one.

As we were saying our goodnights, I heard my name on the loudspeaker, summoning me to the captain's quarters. How nice, I thought; he wanted to say hi and get acquainted. When we arrived, the doctor was there too, looking very distressed. Earlier on, Kari had visited him to say hello, explaining that she was thirty weeks pregnant and with a low-lying placenta. She wanted to reassure him that there was nothing to worry about, that she was in great shape. The captain didn't mince words. "Too risky, we will be isolated. If she hemorrhages we can't do anything on board. After consulting with the doctor, I decided that you must leave this ship tonight." Shocked, we tried to resist, telling them the story of the gentleman at dinner who couldn't stand up. "Now, *he*'s someone who shouldn't be on this ship," I said. Kari was fuming. I pleaded with the doctor to speak to Kari's ob-gyn in the United States, who had given her a clean bill of health. "I can talk to her, but only to yell at her," said the doctor. "She is clearly incompetent. I won't change my mind." He smiled in that most annoying passive-aggressive know-it-all way the French are so good at. To make a long story short, at 11 PM we repacked and left our cabin. Under bewildered looks from the other passengers, I carried my hysterical five-year-old down the plank and off the ship. In a few minutes, we were back at Hotel Borg. Our cruise was over before it had started. Complete disaster.

Now what? After many phone calls to the tour operators and hours of furious planning, we decided to turn lemons into lemonade. We would stay in Iceland, rent an SUV, and explore the island on our own. In three days we would rendezvous with the cruise ship at the port of Akureyri in the north-central part of the island. I would then give a long lecture, combining all the content I had prepared for the cruise. The somewhat desperate plan worked beautifully. We ended up having an unforgettable trip, filled with adventures in Iceland's dramatic landscape, free to roam where we wanted, thanks to a rugged Nissan SUV and an English-speaking GPS with a sexy woman's voice we called Lola.

## Primeval Landscape

We spent our first day exploring the so-called Golden Circle, which is east of Reykjavík and has some of the most famous attractions on the island. The first stop was Geysir, or the Geysir Geothermal Field, which, as the name says, is a large geyser, surrounded by a few smaller ones. It shoots jets of steaming hot water up to about one hundred feet every few minutes or so. A crowd stands around the site, patiently waiting until the jet comes, spewing hot water and steam from underground. Typically, geysers occur in active volcanic regions, where surface water runs underground until it hits hot rocks at depths of roughly one mile. Under fast heating, the water expands explosively upward, shooting to the surface like a steam volcano. In the case of Geysir and its surroundings, the many active geysers lie along a tectonic fault, a deep crack in Earth's crust where rock plates move in different directions.

Iceland is literally being torn apart along the line joining the North American and Eurasian tectonic plates. That's where most of Iceland's fearsome volcanoes lie, acting as relief vents for the furious underground activity. Our goal was to go to Thingvellir, a national park where the fault line is plainly visible from the surface, in the shape of a huge canyon. On the way there we stopped at the stunning Gullfoss, our first of many of the hundreds of majestic waterfalls that punctuate Iceland's rivers. Thunderous waters, crashing sideways into a steep rock wall, make it impossible to see the bottom, giving the impression that there is none, that the fall is a gargantuan hole to the center of the Earth. Although the raw beauty of the sight was entrancing, it didn't stop me from wondering about the fish hidden in those waters. With time passing fast, we decided to take a "shortcut" to Thingvellir on road F338, so that we could also take a good look at Langjökull, the second-largest glacier in Iceland.

It was the road from hell. Some forty miles of awful terrain, very bumpy, including a few nerve-racking river fordings, where tire tracks were visible on our side and the far side, with only fast-moving water in between. Apart from very scattered lichen on rocks, there were no signs of life. We might as well have been on Mars. Kari was bouncing all over her seat, holding on to her (very large) belly so tightly that her arms cramped. I wondered (in silence) if we were going to have a premature Icelandic baby. (Interestingly, our son Gabriel does look like a little Viking.) Amid thick clouds of dust and potholes capable of hiding a yeti, we thought longingly of the huge cruise ship, peacefully cutting through the waters of the North Atlantic. But we persevered and did catch some wonderful distant views of the glacier, and of an amazing frozen desert that preceded it, blending sand, snow, and rocky lava in otherworldly contours. Thankfully, the car and Kari held up. We now understood why the locals say you shouldn't venture into Iceland's remote interior without knowing where you are going. We reached Thingvellir exhausted but safe.

The place—indeed most of the country—feels primal, as if it's still in the early stages of its geological formation, which it is, given that the island emerged from the ocean only about eighteen million years ago. To witness the fault lines, the geysers, the volcanoes, the sulfuric muck oozing from ground holes scattered across the countryside is to take a trip back to a time when furious forces were shaping the young Earth itself. The barren landscape, the abrupt ice-capped fjords, evoke the distant past, before humans walked the Earth. Hiking in secluded places, up the Eldborg volcano to peek over its perfectly shaped cone mouth into the depths, felt as if we were breaking into sacred ground, the realm of Odin and his son Thor. I looked around half expecting to see them there, divinely detached from the realities of men.

To get to my rendezvous with the *Boreal* in Akureyri, we circled the island clockwise from south to north. This gave us an opportunity to explore some less-traveled areas, the fjords around the town of Ísafjörður on the northwestern tip of the island. There is something quite ominous about how the huge elevations tower over the small peninsular town, giants patiently awaiting the right moment to wipe humans out to sea. But the Ísafjörðurians are a resilient bunch and would probably welcome the challenge. Witness their proud hosting of the European Swamp Soccer Championships, an impossible soccer game played on fields with mud up to the knees. Not something many Brazilians would venture into, especially at 66° latitude. Still, even mud soccer is nothing compared to what defines a "true Icelandic man": to swim naked to the island of Drangey, holding a torch in one hand while singing the national anthem. If this is a ritual every teenager must go through, it explains why the population has been dwindling.

Akureyri is a charming city bolted in within what looks like a divot taken out by Odin's bad golf stroke. The result is an elongated bay in the shape of a U, with the town at the bottom. The *Boreal* was already there, an overgrown sore thumb in the town's quaint harbor area.

Kari and I made sure we saluted the captain and the doctor. Awkward smiles were exchanged, as the doctor saw Kari's protruding belly with a mix of awe and disappointment. "We had a great time hiking up a few volcanoes," she said with an impish smile. "Too bad you were stuck on board."

I was conducted to an auditorium, where my "students" awaited. I was excited to finally do my job, talk about our origins, Earth's geological history, and the environment.

## Modern Creation Narrative:
### An Integrated View

Every culture has a creation story, a narrative that attempts to make sense of that most fundamental of questions, Why are we here? In the West we are more familiar with the Bible, where creation is the act of a willful God. The book of Genesis relates a sequence of events of growing complexity as days go by. Quite interestingly, even for almighty, all-knowing God, things don't go as planned; Adam and Eve fumble and eat the forbidden fruit from the tree of knowledge of good and evil. Their harsh punishment — expulsion from paradise, mortality, pain in child labor, working for a living — shows that God wasn't too keen on having his creatures go beyond a certain level of knowledge. Thankfully, our first woman wouldn't accept this limitation and decided to go forward with the first challenge to authority.

Creation myths focus on the hardest question we can ask, that of the beginning of all things. *Creatio ex nihilo*, something from nothing, is possible only for powers that transcend the natural world. It makes sense: only a power that exists outside space, time, and matter can create space, time, and matter. Thus, all religions solve the issue of the first cause by assuming the existence of an absolute power that transcends the laws of Nature.

As products of their originating culture, creation myths vary from place to place. Despite their rich diversity, many share similar attributes, and all can be grouped into two major classes, depending on how they deal with the nature of time: Is the world eternal, or was it created at a certain time in the distant past? Myths that declare the world eternal consider time to be a local attribute, important for men and beasts but not for the cosmos at large. There is either no creation event, as is the case with the Jains of India ("Know that the world is uncreated, as time itself is, without beginning or end"), or the Universe may go through infinitely many cycles of creation and destruction, phoenixlike, as illustrated in this Hindu narrative: "In the night of Brahman, Nature is inert, and cannot dance till Shiva wills it. He rises from His rapture and, dancing, sends through inert matter pulsing waves of awakening sound, and lo! Matter also dances, appearing as a glory round about Him. Dancing, He sustains its manifold phenomena. In the fullness of time, still dancing, He destroys all forms and names by fire and gives new rest." The Hindu god Shiva dances to create and destroy the world, giving matter its vibrant rhythm through each cycle of cosmic existence.*

Myths with a specific creation event are far more common, the Bible being a familiar example. Time has a clear beginning that also marks the birthing of the world and, soon after, of animals and people. The unifying cultural myth of Iceland is called the Edda, written in Old Norse in the early thirteenth century by the Icelandic scholar Snorri Sturluson, incorporating many elements of earlier Viking sagas. Here are some excerpts, illustrating how gods give rise to life:

> The first world to exist was Muspell, a place of light and heat whose flames are so hot that those who are not native to that land cannot endure it.
>
> Surt sits at Muspell's border, guarding the land with a flaming sword. At the end of the world he will vanquish all the gods and burn the whole world with fire.
>
> Beyond Muspell lay the great and yawning void named Ginnungagap, and beyond Ginnungagap lay the dark, cold realm of Niflheim. Ice, frost, wind, rain, and heavy cold emanated from Niflheim, meeting in Ginnungagap the heat, light, and soft air from Muspell.
>
> Where heat and cold met appeared thawing drops, and

---

*In my book *The Dancing Universe: From Creation Myths to the Big Bang,* I offer a more detailed discussion of creation myths and their relation to modern cosmological models.

this running fluid grew into a giant frost ogre named Ymir.
. . .

   Odin, Vili, and Vé killed Ymir and out of him made the
sea and lakes [blood], the earth [skin], trees [hair], mountains [bones]. Maggots in his flesh were transformed into
men. Skull [sky]; clouds [brains].

So, according to the Edda, we are all descendants of the maggots
that ate the flesh of the slain giant frost ogre Ymir — not a very
charming picture of our origins.

   Now, compare this Icelandic mythic narrative with the Chinese myth of P'an Ku, written around the third century CE. According to the Chinese, at the beginning of time, the Universe
was a cosmic egg. When the egg broke, the giant P'an Ku came
out of it, together with the two basic elements, yin and yang. But
that was only the beginning: "The world was never finished until
P'an Ku died. Only his death could perfect the Universe: from
his skull was shaped the dome of the sky, and from his flesh was
formed the soil of the fields; from his bones came the rocks, from
blood the rivers and seas; from his hair came all vegetation. His
breath was the wind, his voice made thunder; his right eye became the moon, his left eye the sun. From his saliva or sweat came
rain. And from the vermin that covered his body came forth man-
kind." How could two cultures separated by such enormous distance and ten centuries come up with pictures of creation based
on the metamorphosis of a dying divine being into the world, including our origin from vermin? Cultural anthropologists and
religion scholars can spend fulfilling academic careers studying
comparative mythology, identifying common trends and also
differences among the myths of many cultures. Clearly, there are
a few universal symbols of creation that recur throughout time,

what Jungians could call *archetypes of creation*: there are only so many ways that the human brain can contemplate the notion of the origin of all things, including its own.

Moving on to modern science, we have now, after a century of false starts and phenomenal discoveries, constructed our own creation narrative based on solid data and observational evidence. Remarkably, for about five decades or so before we had data about the early Universe, the various cosmological models repeated some of the same archetypes of creation found in myths: the eternal Universe, the phoenix Universe, the creatio ex nihilo Universe, and so forth. We now have a commonly accepted narrative, based on Big Bang cosmology, wherein time did have a beginning, at least in our cosmic patch of space, about 13.8 billion years ago.*

Within our Universe, space and time emerged together with matter. So the question, What happened before the Big Bang? doesn't really make much sense, unless the Big Bang wasn't the beginning, a speculation we leave for another time. As matter evolved into ever more complex arrangements, time ticked away and space stretched. We are unclear as to what happened within the first fractions of a second after the "bang," although we have a firm grip as to what happened from about 0.01 second onward. We can *speculate* about what happened much earlier, but I will

---

*There are nuances that I am leaving out. Some models, as we have seen, call for a collection of universes called the multiverse. In principle, the multiverse is timeless; time becomes a "local" variable, beginning to tick when a baby universe comes into existence. Other ideas, including a recent paper of mine with Stephon Alexander and a graduate student, Sam Cormack, call for a revival of the phoenix Universe, wherein the fundamental constants of Nature have slightly different values in each cosmic cycle. Although each cycle has its own creation narrative and specific time, taken together our phoenix Universe is timeless and uncreated.

skip that here for brevity. Between a second and three minutes, protons and neutrons joined to make the nuclei of the lightest chemical elements, hydrogen, helium, lithium, and some of their isotopes. At about 380,000 years, electrons joined protons to make the first hydrogen atoms. A few hundred million years after that, clouds of hydrogen collapsed to make the first stars, enormous giants that lived only fleetingly (by stellar standards) before spewing out their entrails and many heavier chemical elements across interstellar space. In a billion years, the first galaxies were forming, collecting millions, even billions of young stars. As stars were born, so were their planetary consorts, circling around as if to admire the shining beauty at the center of their orbits.

Some four billion years after the bang, our Milky Way galaxy emerged from groupings of smaller galaxies. Five billion years after that, the solar system emerged from a collapsing hydrogen cloud thirty thousand light-years away from the galactic center. There, in the third planet from the Sun, from the sludge accumulated for hundreds of millions of years, the first life forms pulsed and floated around, unaware of the great adventure that was about to happen. For three billion years, little happened, but certainly an important little, as these first creatures mutated into photosynthetic algae that generated energy by changing $CO_2$ into oxygen. Now in a richly oxygenated atmosphere, and with unicellular creatures boasting a more evolved genetic material, life exploded in amazing diversity to populate the oceans, land, and air. Out of this cacophony of living entities, struggling for food and survival, some slowly evolved to become giant cold-blooded reptiles about two hundred million years ago. Then, about sixty-five million years ago, a giant asteroid came crashing down from the sky, wiping the dinosaurs out of existence and changing the planet in dramatic ways. Life endured, and out of the survivors new kinds of species evolved, including those that would become our primate ancestors. Humans appeared only very recently, about two hundred thousand years ago, competing for space and resources with other humanoids. We prevailed, thanks to our thicker frontal cortex, it seems, and out of our dominance came civilization as we know it. The rest, as they say, is history, quite literally.

The fundamental lesson from our modern creation narrative is the essential unity of all that exists. Everything came from the same primeval fireball we call the Big Bang. Every star, planet, moon, every living creature, shares the same chemical elements, the same cosmic dust that fills the emptiness of space. Every living creature is animated stardust. Every creature now living on this planet shares a common ancestor, a being that existed billions of years ago. Science tells us that there was material unity in the past; all that exists has a common origin. Time has split this unity, as the diversity we see around us has unfolded. Being the only known creatures aware of our joint beginnings, we have the moral duty to respect the diversity of life. More than a moral duty, it is something we should celebrate, knowing that each and every being carries with it the seed of who we are.

This is true for the food we eat and for the trout we fish. Such reflections, coupled with the brutal treatment of animals in meat plants and the global environmental impact of industrial livestock and poultry production, made me into a vegetarian many years ago. We all have to eat, of course, and meat is a great source of protein. But using cattle to convert vegetable protein (what cows eat) into animal protein (what we eat) is not efficient. Millions of acres of forest are being cut down annually to open fields for pasture. Water is diverted from rivers and from agricultural farming to irrigate cornfields that feed cattle. Methane gas from cattle contributes to global warming in increasing amounts worldwide.

Brazil and the United States are the largest meat producers in the world. JBS, a Brazilian company headquartered in São Paulo, leads Tyson Foods in the United States. According to a report based on figures from the United Nations' Food and Agriculture Organization by Magda Stoczkiewicz, the director of Friends of Earth Europe, "nothing epitomizes what is wrong with our food and farming more than the livestock sector and the quest for cheap and plentiful meat." The disturbing report, called *Meat Atlas*, can be found on the Internet.

The numbers are truly horrifying. The *daily* slaughter count from JBS alone amounts to twelve million birds, eighty-five thousand head of cattle, and seventy thousand pigs, sold to 150 countries. Animals are squeezed into cramped spaces, cruelly treated, and brutally killed. To increase profit margins, workers receive low wages and bad compensation packages, increasing their stress at work. To avoid disease, cattle are fed huge amounts of antibiotics: about fourth-fifths of all antibiotics sold in the United States go to livestock and poultry. This overuse results in antibiotic-resistant "superbugs," which pose a mounting health threat. Natural selection will inexorably lead to drug-resistant mutations in bacteria. Over twenty-three thousand people die annually in the United States from antibiotic-resistant infections.

There is a long — mostly unknown or conveniently ignored — chain of events from the cute calf being born at the barn to the neatly packaged meat people buy at the supermarket. Yes, meat tastes good and people have to eat, all eight billion of them. But we have to start asking ourselves how long we are going to ignore what is obvious — that our meat-eating culture is not environmentally sustainable and is morally abhorrent. We've moved far from our Paleolithic ancestors. It's time our diet followed our cultural advances.

Whenever I bring this topic up, I try hard not to sound too condemning or judgmental. After all, I ate meat for a long time. Even if I resisted it as a child, disgusted by all the nerves and fatty bits, I was forced early into the meat-eating culture, as every other Brazilian child still is. Many tears were shed. After chewing and chewing the same chunk of meat, unable to swallow it, I'd run to the bathroom to spit the gross gray mass into the toilet, pretending I had to pee. I have vivid memories of walking down the streets and watching, perplexed, the huge flanks of meat hanging from metal hooks at the butchers' shop, blood dripping onto fly-infested puddles on the floor. The vision still makes me shudder. How could people eat that? But we did, and do, and will, even though there are other sources of protein out there, equally effective, definitely healthier, definitely more humanely produced.

Critics of vegetarianism argue that Nature is ruthless, that man, at the apex of the food chain, has earned the right to do as he pleases with other species. Predators eat meat, they claim; the law of the jungle and of the seas is the law of the strongest, the law of blood. True enough; but we are not lions or sharks. We no longer need to kill other animals to survive, as lions and sharks do. Even if large sectors of the world population still depend on meat for survival, others who can avoid it should, given what we now know. Thankfully, the difference between us and natural predators is becoming clearer as more people move toward sustainable farming and free-ranging, antibiotic-free, and more humane treatment of cattle, pigs, and poultry. ("Humane" is a curious adjective, as it implies that being human means, in essence, being good.) Seafood opens another huge can of worms, as commercial fishing vessels the size of football fields are quickly depleting the oceans. But I will leave fishing for later, when I explore the unpleasant question of how recreational fishing (including fly-fishing) affects the sustainability of ocean life.*

## Planetary Awareness

My lecture onboard the *Boreal* continued with an overview of global warming. It's hard not to think about this issue when we are in Iceland. Early in 2015, a team of scientists from the University of Arizona published a study showing that the island is growing, its landmass rising at an alarming rate of 1.4 inches per year. The reason, the scientists demonstrate, is the fast melting of glaciers, which reduces the pressure on the ground, allowing for land to spring upward. To reach this conclusion, the scientists analyzed twenty years of GPS data from more than five dozen points around the country. The numbers are alarming: the island loses about eleven billion tons of ice per year, in a process that tends to accelerate as the ice mass decreases. If the process continues at its current clip, by 2025 the island could grow as fast as 1.6 inches per year, about the same average rate as a first grader. As the island rises, the pressure release may cause rocks deep underground to melt, fueling the already eager Icelandic volcanoes. The prognosis is not encouraging.

When thinking about global warming, there are two essential questions to consider: First, is the average global temperature increasing over time? Second, if it is, is the increase caused by human interference, or is it attributable to natural sources? The es-

---

*For a broader canvas, the reader can consult Paul Greenberg's book *Four Fish: The Future of the Last Wild Food.*

sential idea is disarmingly simple: if more heat hits the Earth than goes back out to space, the Earth heats up; otherwise, if the Earth gives up more heat than it receives, it cools off. Global warming occurs when the Earth heats up. The hard part is to compute the details of this heat flow in time, including all the potential sources that can interfere with it. That's where climate research comes in, mathematical models that attempt to include all the relevant factors influencing the climate and extrapolate the results to the future, to predict what comes next, short and long term.

We can think of the atmosphere as a sort of thin blanket around the Earth. It's an interesting kind of blanket, as it can let some kinds of radiation in and others out. A normal blanket functions as a heat barrier, keeping some of the heat our body generates close to our skin. A thicker blanket makes it harder for our body heat to dissipate outward. In the case of the atmosphere, most of the heat comes from the outside, from the Sun. There are, of course, terrestrial sources of heat, such as natural radioactivity (elements that decay spontaneously, releasing energy in the form of particles) and volcanic activity, generally grouped together as "geothermal heat flux." But these natural terrestrial sources are subdominant (less than 0.03 percent) when compared with our celestial provider of radiation. Most of the radiating power from the Sun gets to the upper atmosphere in a combination of three kinds of radiation: visible (40 percent), infrared (50 percent), and ultraviolet (10 percent). Thanks to the atmosphere's protection, at the surface ultraviolet light goes down to 3 percent, while visible light goes up to 44 percent. Eventually, once some of the radiation is absorbed, some is reflected, and some is used up, part of it bouncing back to outer space in degraded form, mostly infrared. We can think of the Earth as an engine that churns high-grade solar radiation into low-grade infrared radiation. Global warming occurs when certain gases in the atmosphere retain more radiation, effectively working to thicken the blanket. Some of this radiation is reflected back to the surface, causing the temperature to increase.

An opposite, cooling effect occurs when dust particles are suspended in the atmosphere (aerosols). The dust blocks some of the sunlight, and the temperature goes down. A dramatic example of such cooling occurred when a seven-mile-wide asteroid hit the Yucatán Peninsula in Mexico sixty-five million years ago. As we have seen, this impact is considered one of the main culprits for the extinction of the dinosaurs, together with about 50 percent of all life. The impact lifted a huge amount of dust into the atmosphere, causing a long dark winter that lasted for decades. Less dramatically, volcanic eruptions can spew enough dust into the atmosphere to cause changes in temperatures, sometimes globally. In 1883, the Indonesian Krakatoa volcano erupted with huge fury, affecting the global climate for about four years. The winter of 1887–88 had powerful blizzards, and record snowfalls were seen worldwide.

For decades, scientists around the world have been comparing their climate change models with increasingly more accurate data collected from weather stations covering most of Earth's surface and lower atmosphere. Results are carefully analyzed and discussed among hundreds of experts, and regularly tabulated in IPCC reports (IPCC stands for International Panel on Climate Change). Here is how our two main questions are currently answered: First, there is no question that taken over the past 150 years, the average global temperature is slowly rising. There are normal year-to-year fluctuations, but the average trend upward is uncontroversial. Since industrialization took off at the beginning of the twentieth century, the increase has been steady, when averaged over a five-year interval. More distressingly, the increase really takes off around 1980 and is not showing signs of abating. Only a very small minority of scientists and those motivated by political or personal agendas question this. Without going into the details, the language of the IPCC reports has been changing from moderate to practically certain. The planet is getting warmer. In particular, and taking all errors into account, the data link 2005, 2010, and 2014 as the three hottest years on record. Ignoring this fact is like stepping in front of a train and hoping that if we close our eyes the train won't hit us.

The second question, whether human activity is the main culprit in global warming, is harder to consider and certainly more controversial, given that its answer affects economic and political decisions. Economics and politics aside, however, it is primarily a scientific question. One needs to isolate different possible causes for warming and quantitatively analyze their local and global impact. Again, scientists are nearly unanimous in relating the rising temperatures to an increase in pollution, in particular due to deforestation and the burning of fossil fuels. Accumulation of the so-called greenhouse gases, those that increase the atmosphere's reflectivity (thickening the blanket), is to blame, the main contributors being carbon dioxide, water vapor, methane, nitrous oxide, and ozone. On the one hand, if none of these gases were present in the atmosphere, the Earth would be cooler by fifty-nine degrees Fahrenheit. On the other hand, too much of them leads to overheating.

Since the beginning of the Industrial Revolution, the amount of carbon dioxide in the atmosphere has grown from 289 parts per million (ppm) to an alarming 398 ppm in 2014, according to data collected at the Mauna Loa Observatory in Hawaii. Climate models indicate that by the time carbon dioxide concentration is double that of early Industrial Revolution levels, the global temperature may rise between 3.6 and 9 degrees Fahrenheit (2°C–5°C). Because of the statistical nature of climate modeling, it's hard to assign an exact date to when this will happen. However, all models indicate that unless a radical cut in carbon emission happens within the next few years, this temperature rise will take place before century's end. Changes are already happening, even if specific local events are harder to pinpoint. (For example, we can't say with certainty that a rise in tornadoes in the Midwest during a certain month is directly related to global warming, although weather instability is definitely a consequence. In fact, although the number of tornadoes is lower now than it was sixty years ago, tornado clusters — many in sequence in a short period of time — are on the rise.)

Putting together the findings from modern astronomy and climate modeling from the past few decades, we arrive at an inescapable conclusion: Earth is a rare and finite planet, with finite resources. We evolved here and depend on its long-term stability for our survival. With almost eight billion people to feed and house, we need to develop a more sustainable relationship with our home planet, the only one that can support us for the foreseeable future.

Awareness of climate change and planetary sustainability faces two major challenges: first, changes happen gradually, too slowly to alarm most people. Second, there is enormous political and economic pressure to maintain the status quo. Alarmism, describing doomsday scenarios of major social unrest and global upheaval, does little to change people's minds. Realistic change, a change in mind-set, will happen only as new technologies based on renewable energy sources become economically competitive or, even better, economically attractive: when a combination of solar, wind, and biofuel power generates cheaper electricity than coal, gas, and oil. This is a technical challenge, one that can be met within the next two decades or so. The worth of a generation is measured by its legacy. I hope, for our children's sake, that my generation's legacy will be a better planet, populated by more spiritually attuned humans.

## The Temple

Lecture completed, it was time to drive back to Reykjavík so that Kari and Lucian could fly to the United States and my fly-fishing adventure could finally start. I had arranged everything through the Reykjavík Angling Club with the very competent Haraldur Eiríksson. Ari, my guide, would be waiting for me at the Akureyri airport and drive me straight to the Hof, a small inn for Laxá River fly-fishermen. Any other kind of fishing around these parts is akin to heresy. I had two full days at the Laxá, considered one of the best brown trout rivers in the world. The river runs from Lake Mývatn, a spectacular bird sanctuary with plenty of food flowing downstream. The result is a large population of overfed brown trout, the average being around five pounds, some huge ones reaching eleven pounds and more. A very different story from any trout fishing I had ever done.

Along the drive from the airport, Ari explained that fishing is from 8 AM to 2 PM, and 4 PM to 10 PM. That's a twelve-hour day of fly-fishing, more than enough for even the highly addicted. The wonderful thing about fishing at the Laxá in July is that you go after giant trout with dry flies. No fly-fishing experience compares with catching a large aggressive trout on a dry fly. Although I couldn't wait to try my luck, a strange uneasiness started to creep in as we approached our destination.

The Hof is a very simple place, about one hundred kilometers from Akureyri. Small bedrooms with a tiny toilet, a communal shower, and a dining room. People don't come here for comfort, although the food is very good, if a bit trying for a vegetarian. The potatoes and vegetables from local farms were my salvation.

As we arrived in late afternoon, I had time to hike around and get to know the area. It was cooler than I expected, and I had to bundle up in a few layers. The Laxá's dark blue water flows into endless pools and small waterfalls, its many branches, broad in some spots and narrow in others, offering a seemingly infinite array of possibilities. It's not a river but a collection of waterways, meandering and bifurcating in chaotic fashion along a very wide plain. In the distance, I could see a few signature Icelandic white farm buildings with red roofs, surrounded by majestic snow-capped peaks. Moss and colorful lichen were everywhere, decorating the ancient lava fields, softening the desolation of time and inclement weather. The riverbank was easily accessible, as the vegetation along the water is low — a welcome break from the ever-present trees and brambles of New England. It quickly became obvious that without a guide, one could spend a year exploring the river, wondering where to go. Wading can be tricky, as the current varies from spot to spot, sometimes quite abruptly. I noticed a few trout rising after a cloud of mayflies. They seemed huge, more like salmon. My heart quickened with a mix of anticipation and weariness.

I found a place to sit, to take it all in. A long time, seven years, had passed since that first walk along the Dartmouth Green, rods waving in the air to Rick's methodic counting. From my early cinematic Brad Pitt take on fly-fishing to sitting at the margins of the Laxá, I had embraced the sport and made it my own, struggling to learn its artful ways as best I could. I was not a master, like Jeremy or Luca Castellani, but that was never my intent. I learned much from them, and have grown as a result. Grown as a fisherman, and grown closer to Nature, which had always been my main motivation. I never fished for the fish. If I caught one, great, I would admire its beauty and hurry to return it to its watery world. The truth is that I always felt a twinge of guilt when I looked at the struggling fish drowning in all that air, staring at me with startled eyes. Even in graceful fly-fishing there is violence, the forceful taking of a being out of its natural environment. Sitting along the Laxá, a dream place for any fly-fisherman, I started wondering what would come next.

Since I had been that young boy of eleven, fishing had been a portal into a spiritual dimension of being, a way to transcend the clutch of time. Waving the rod, stepping into the current, watching the line floating along, senses sharpened, eyes riveted on each and every change — fishing has always been an attempt to establish a relation with a dimension beyond my own. To me, fly-fishing is a form of meditation, of letting the self go, a way of approaching an emptiness of being that is its fullest expression. By doing you don't do, by acting you don't act. With a shudder, I realized that catching a fish, the end result of the act, was an interruption. The electrifying hit, the rod bending, the line whooshing away, the adrenaline spike, were alarms bringing me back to the present, to the ever-so-real passing of time. Paradoxically, catching the fish spoiled my experience of fishing, my search for timelessness.

I heard Ari calling from the Hof, inviting me to join him for dinner. I walked back slowly, shaken to the core by my epiphany. I couldn't share this with anyone yet, certainly not with the other fishermen at the Hof, who would consider me either crazy or a complete fool. In good Icelandic fashion, some of them were sitting in a Jacuzzi, drinking wine and eating herring. Among them was Gunnar Egilsson, who had built a monster truck to break the speed record for driving over Antarctica's icecap. His wife, a solidly built blond lady in her late forties, saw me pacing around, somewhat lost.

"C'mon, you can join us! We look awful but we don't bite," she said. Gunnar smiled approvingly.

I changed into a bathing suit and hopped in, feeling very awkward.

"So, what are you doing here? Running from police?" General laughter. She did all the talking, apparently.

"Here, have some wine with us. It's horrible, Chilean. If it were Argentinian, there wouldn't be any left for you!"

I looked at the label, a perfectly fine bottle of Carménère. I thanked them profusely and just sat there, sipping wine from the other side of the planet, listening to that strangest of strange languages, trying unsuccessfully to pick up a familiar word or common root here and there. The chef stepped outside and yelled something.

"We eat in ten minutes. Hurry up or she won't feed you!"

When I arrived at the dining room, Ari had left a message saying he had to go home. It was about 10:45 PM, a normal time for dinner here, when the fisherman are returning from the river and the sun is beginning to slowly arc its way underneath the horizon.

"What are you doing sitting alone? Come join us!" Gunnar's wife, Edda, would not have it any other way. She grabbed my fork and knife and placed them at their table. She was not someone to trifle with.

"So, how was fishing?" she asked. I told her I would be starting tomorrow. "You will have a great time!" They came every year, she told me, around mid-July. Gunnar asked me what I did and where I came from. They were intrigued by the idea of a Brazilian-born theoretical physicist working in the United States. She saw my plate and shook her head. "So, that's why you don't eat meat, right? From principles."

"Very observant," I answered.

"For Gunnar's fiftieth, we had 250 people, and we all ate fish and meat we killed ourselves!" Gunnar smiled approvingly again. "So, you eat fish you catch?"

"No, I put it back."

"So why do you fish then?"

"Well, I . . ."

"For my fiftieth, Gunnar and I go to Alaska! I want to fish salmon and hunt bear!" I pictured Edda wearing a Viking helmet, holding a sword with one hand and a decapitated bear's head with the other. I smiled, wondering quietly what drove people to be such proud killers of innocent animals. What right do we have to kill living creatures with such impunity and even admiration? Unless it is for survival, what gives us the moral right to determine another living being's fate?

I went back to my room to organize my gear for the next day and try to sleep. When I opened my window, my jaw dropped. The sky was on fire, the sun right below the horizon lighting up a sea of cloudy waves. It was the most amazing sunset I'd ever seen, not least because it persisted for more than an hour. There is no night to speak of in July at those latitudes, and I had to force myself to close the window shades and try to sleep. After all, the next day would be fly-fishing in paradise.

I was happy to have delicious Icelandic *skyr*, a rich yogurt topped with granola and honey, and strong coffee. It was colder out than I expected, so again I had to bundle up for the day's adventure. Ari was waiting for me promptly at 8 AM, and we

quickly took off to the river. A fifteen-minute walk put us at a narrow branch, no more than thirty yards across. The current was
slow and steady, the water deep and dense looking. The different fishing spots (they are called "beats") have names, but they
all blurred in my head after a few hours. This one was not too far
from the road, although the water deafened any car noise from
afar. Ari set my 8-weight rod with a black-gnat fly and told me to
cast upstream and let the fly drift toward me, "British style." The
rod bent in half almost immediately, as the fifteen-inch brownie
struggled for freedom. I had learned my lessons and let the fish
roam about without forcing it my way, keeping the fly line tight
and the rod upright all along. Whenever the fish stopped to rest, I
brought it a bit closer, ever so gently. Yesterday's insight was playing its role, as I tried very hard to minimize the tug on the fish's
jaw without prolonging the fight. Perhaps there is a road to com-
passionate fishing? The absurdity of the thought made me smile.
After some five minutes of give and take, I had my first gorgeous
Icelandic trout, golden brown with dark brown spots along its
body. I could plainly see why trout belong to the family of salmo-
nids. Remove the spots and it could easily pass for a salmon, at
least for the less clinical eye.

After two hours in the same spot, I had caught four trout, ranging in size from fifteen to twenty-two inches, some of the biggest
trout I had ever seen, or thought existed out there. Excited, Ari
took me to a different spot, a large, slow-moving section of the
river where he promised the "dinosaurs" lived. By dinosaurs he
meant enormous trout, some reaching thirteen pounds. A few
minutes after we got there, I could see dorsal fins cruising up and
down, the water twirling as the fish moved their massive bodies
this way and that after small caddis flies. I cast and I cast and I
cast in every possible direction: with the current, against, at dif-

ferent angles — and nothing. We tried different flies, different sizes of flies — and nothing. The fins kept cruising along, tantalizingly close and yet completely indifferent to our attempts. Those were older fish, masters of their whereabouts, wise and weary of human interference. Our trickery would not fool them.

We broke for lunch and a much-needed rest. We still had six hours of fishing in the afternoon, exploring different parts of the river. Ari was a quiet guide, who basically took me to the spots he knew and watched me from a distance, camera in hand. Only once or twice he advised me in the direction and method of casting. A connoisseur of these waters, he chose all the flies. The wind was mild and the weather quiet. Conditions were perfect. We spent the next few hours following the river as it meandered along the plain, stepping into the water at different places. The fishing was the best I'd ever experienced, in both the size and the number of fish. I felt accomplished and somewhat relieved when we were done. Catching the fish was weighing on me. Heraclitus understood this long ago, when he proclaimed it impossible to discover the limits of the soul, such are the depths of its meaning. Sometimes we set out on a quest only to find out midway that our goals have changed. But I still had another day at the Laxá, which promised to be as good as the first.

At dinner, Edda was eager to hear my adventures. She saw me getting some meat lasagna, out of desperation.

"So, eating meat already, huh?"

"Yes, need the protein," I replied, embarrassed and not too happy with myself. I have learned since then to always travel with my stash of nuts and protein bars. Edda broke into a huge smile.

"Now you are a real man!" She said triumphantly. The whole room burst out laughing. Even Thorodur E., the distinguished and exceedingly quiet gentleman wearing a Sherlock Holmes hat and smoking a pipe, a member of the "river committee" and legendary fly-fisherman, managed to crack a discreet smile. "Next year you will eat bear meat with us!"

I crashed into a deep sleep, preparing for the next day, my last. Ari drove me to a different part of the river, where a sequence of low-lying waterfalls worked to oxygenate the water, bringing nutrients downstream: a picture-perfect spot for fly-fishing, with alternating shallow and deep pools along the riverbanks. Easy entrance into the water, and not too hard wading along, if you have boots with good grips, especially now that they can have only rubber spikes. It was a magical day—no wind—cloudy and fairly warm, at least for Iceland. Everything seemed to click into a primal rhythm: the water flowing along, my casting, my half-submerged body balancing on the rocks, my line management. I heard the voices of all my instructors and guides over the years, gentle men in love with the outdoors, with the art of fly-fishing. To them, too, fishing was a portal into a different reality, an opening of a window toward a hidden part of the self.

The hours passed in a minute, as if time had ceased to be. The wild trout graced me with their visits, amazing creatures that they are, in such perfect harmony with their environs. I thought of the lines by William Blake,

> Tyger Tyger, burning bright,
> In the forests of the night;
> What immortal hand or eye,
> Could frame thy fearful symmetry?

as I admired the imperfect symmetry of each trout I caught and hastened to release, their strong golden bodies reflecting the sinking sun. There was something sacred in each of them, in their graceful design, a product of millions of years of evolution, of trial and error in the game of life. If we could only live as close to our environment as they do so effortlessly, never taking more than is their due, respecting the rhythms of their water world, conserving energy at all times, we would redefine the way we inhabit this planet and deal with its resources. The trout have much to teach us, if we are willing to listen.

I told Ari I wanted to spend my last hour alone, fishing close to the Hof. He gracefully agreed, dropping me with a handful of different flies at the first spot I tried out. I entered the water knowing where to go. The trout were where I thought they would be, the most reasonable of places, where the current splits behind a boulder and slows down on one of the two forks. Hardly any swimming needed, food drifting right into their eager mouths. I cast upstream and let my dry fly float along with the current, hoping for the simple beauty of the unexpected. No fisherman knows if and when the fish will bite; surprise is unavoidable, as is the adrenaline rush of seeing the rod bend, feeling the pressure of the fly line pushing against your fingers.

As I brought a small but beautiful trout close to me, the boy appeared, floating on the water. He smiled quietly, knowing how I felt, happy and conflicted. I showed him the fish, proud as he had been when he caught that huge one at Copacabana beach, over four decades before.

"Quickly, put it back in the water," he said. And as I did so, I could feel his warmth near me, next to me, in me.

I ate alone that night. Gunnar and Edda were gone, and the place seemed deserted. Before going to sleep I stepped outside once more, to hear the river and watch the sun hide below the hills. My thoughts drifted to more immediate concerns, related to how sustainable sport fishing is. Apart from selective spots on the planet such as the Laxá River, no native trout population survives intact. In the United States, some thirty-eight million people buy fishing licenses per year, eight million of them to catch trout and salmon.* To support this fishing industry, federal and state wildlife agencies "stock" around 130 million trout in rivers and lakes across the country. Studies show that the survival rate of stocked fish is below 30 percent. Worse, to feed a yearly hatchery production of twenty-eight million pounds of trout takes about thirty-four million pounds of feed, mostly pellets produced from herring, menhaden, and anchovies harvested from already depleted marine fishing grounds. The waste from the hatcheries is usually dumped untreated into nearby waters. Combined with the introduction of nonnative species to sustain the fishing pressure, what is an artful sport becomes another environmental predation.

⊚    ⊚    ⊚

Something had changed inside, a feeling of complicity with the fish, of humility as a fellow living creature sharing the same planet. We all struggle for survival, animals and men. Nature knows not what kindness is. Nature knows not. But we humans know, or should know, even calling ourselves *Homo sapiens*, literally "wise man." How wise have we been, sharing this world with other creatures? How wise is it to shoot a lion, an elephant, to kill a shark or a salmon for entertainment, for fun, for trophy and bragging rights? What morals guide the finger that pulls the trigger or yanks the fish out of the water? There are many ways to reach a sense of accomplishment, ways that do not call for the taking of innocent lives. We can be close to Nature without maiming its creations.

*The reader can consult *The Quest for the Golden Trout: Environmental Loss and America's Iconic Fish*, by Douglas M. Thompson.

I knew this, but always thought that catch and release was a good compromise, kinder to fish and Nature. It isn't. If catch and release worked to sustain the local trout population in a stream, stocking wouldn't be necessary. The huge resources that go into farming and stocking fish show an implicit imbalance in our relationship with the rivers we love.

As I stood up to go back to my room, I heard a huge splash, probably a ten-pounder going after a mayfly. I recalled Luca's words about night fishing, that it was the purest form of fishing, all instinct. You had to be one with the river, with its sounds, with the trout. At that moment, I realized I could be all that without holding a rod in my hand. The temple, it turns out, is the world.

# ACKNOWLEDGMENTS

This book is, first and foremost, an expression of
my love for the natural world. Over the years, I have been
privileged to experience so much of it through my professional
wanderings around the globe. I'm thankful for having had
the opportunities that I have, and for the support of so
many of my colleagues throughout the years.

I first thank my wife, Kari, for her wisdom and
companionship, for knowing what I need and for knowing
how to tell me. I am blessed by her light every day.

I thank my editor Stephen Hull for his interest in
this project and for his unwavering support of my not-so-
conventional style-breaching narrative. I also thank my agent,
Michael Carlisle, for his always-wise guidance.

Finally, I dedicate the book to all the trout out there that
manage to dodge the constant human intrusion into their world.
Like the equations we cannot solve, their essence will remain
unknown to us, filling us with awe and admiration.